LIVES IN SCIENCE

Lives in Science

HOW INSTITUTIONS AFFECT ACADEMIC CAREERS

Joseph C. Hermanowicz

The University of Chicago Press   *Chicago and London*

The University of Chicago Press, Chicago 60637
The University of Chicago Press, Ltd., London
© 2009 by The University of Chicago
All rights reserved. Published 2009.
Paperback edition 2012
Printed in the United States of America

22 21 20 19 18 17 16 15 14 13 12   2 3 4 5 6

ISBN-13: 978-0-226-32761-7 (cloth)
ISBN-13: 978-0-226-00564-5 (paper)

Library of Congress Cataloging-in-Publication Data

Hermanowicz, Joseph C.
　Lives in science : how institutions affect academic careers / Joseph C. Hermanowicz.
　　　p. cm.
　　Sequel to: The stars are not enough.
　　Includes bibliographical references and index.
　　ISBN-13: 978-0-226-32761-7 (cloth : alk. paper)
　　ISBN-10: 0-226-32761-2 (cloth : alk. paper)
　　　1. Physics — Vocational guidance. 2. Physicists — Social aspects.　I. Hermanowicz, Joseph C. Stars are not enough. II. Title.
　　QC29.H47 2009
　　530.023 — dc22
　　　　　　　　　　　　　　　　　　　　　2008028283

♾ This paper meets the requirements of ANSI/NISO Z39.48-1992 (Permanence of Paper).

In honor of my mother

The career, when success has been achieved, is certainly very pleasant; but the agonies which are endured in the search for that success are often terrible.

ANTHONY TROLLOPE, *An Autobiography*

CONTENTS

List of Tables · ix
Acknowledgments · xi

INTRODUCTION · 1
Guiding Theoretic Perspectives · 4 | Physics and Physicists · 16
Organization of the Book · 19

CHAPTER 1 Following the Scientists · 21
Foundations of the Follow-up Study · 21 | Research Design and Sample · 33
The Ten-Year Career Interval · 39 | The Fieldwork · 45 | Analysis of Data · 51
Academic Worlds Then and Now · 57

CHAPTER 2 Early- to Mid-Career Passages · 76
Professional Profile · 76 | Early Career Patterns · 82 | Elites · 86
Pluralists · 104 | Communitarians · 116 | Summary · 129

CHAPTER 3 Mid- to Late-Career Passages · 138
Professional Profile · 138 | Mid-Career Patterns · 143 | Elites · 146
Pluralists · 155 | Communitarians · 164 | Summary · 171

CHAPTER 4 Late- to Post-Career Passages · 178
Professional Profile · 178 | Late-Career Patterns · 183 | Elites · 187
Pluralists · 198 | Communitarians · 205 | Summary · 211

CHAPTER 5 Lives of Learning · 217
Expectations and the Rhythm of Careers · 218 | Anomie and Adaptation · 221
Reference Groups and Social Control · 227 | Selection of Reference Groups · 234
Rejection of Reference Groups · 237 | Social Control of the Life Course · 240
Careers in Other Academic Fields · 252 | Future Cohorts of Scientists and Contexts
of Science · 260

APPENDIX A Interview Protocol—Foundational Study, 1998 · 269
APPENDIX B Contact Letter to Scientists · 272
APPENDIX C Thank-You Letter to Scientists · 274
APPENDIX D Interview Protocol—Longitudinal Study, 2004 · 275
APPENDIX E Post-Interview Questionnnaire · 279
APPENDIX F Departmental Questionnaire · 281
APPENDIX G Propositions Generated by the Study · 288

Notes · 295
References · 303
Index · 319

TABLES

1. Sample titles and authors of popular books in physics · 18
2. Interview questions on career success and failure · 24
3. Correspondence between departmental identity and institutional identity, by National Research Council rankings · 29
4. Number of scientists, foundational study · 30
5. Institutional classification systems · 32
6. Research design · 35
7. Voluntary and involuntary attrition · 36
8. Scientists who moved to other institutions · 36
9. Institutional types to which scientists moved · 36
10. Graduate institutions of scientists, by current institutional identity · 38
11. Number of scientists, longitudinal study · 40
12. Rank advancement of scientists · 41
13. Age, age ranges, and years of experience of scientists · 44
14. Retired scientists · 45
15. Summary sources of data · 51
16. Conditions of practice at foundational study · 58
17. Conditions of practice at longitudinal study · 62
18. Administrator-reported change in departments · 68
19. Federal spending for academic research · 72
20. Cohort characteristics: The early- to mid-career cohort · 77
21. Early-career patterns of scientists · 84
22. Early- to mid-career patterns of scientists · 130
23. Cohort characteristics: The mid- to late-career cohort · 139
24. Mid-career patterns of scientists · 144
25. Mid- to late-career patterns of scientists · 173
26. Cohort characteristics: The late- to post-career cohort · 179
27. Late-career patterns of scientists · 184

28 Late- to later/post-career patterns of scientists · 212
29 Modal career patterns, overall satisfaction, and work attitudes of scientists, by career phases · 220
30 Definitions of success, recognition sought, career foci, and scientists' objects of satisfaction, by career phases · 230
31 Scientists' perceptions of the reward system of science, orientation to work, and work intensity, by career phases · 242
32 Scientists' perceptions of peak career, professional aspirations, work/family focus, career progress, and attributions of place, by career phases · 244
33 Scientists' perspectives on seeking an academic career again, what scientists would do differently, prominent concerns, thoughts of leaving present institution, and thoughts of retirement, by career phases · 263

ACKNOWLEDGMENTS

To acknowledge is to speak of gifts, most of which cannot be returned in kind. The best that can be done is to record them into print so that our memory of them is eternal, and thus bestow upon the givers of magnificent gifts a kind of immortality. This work was made possible by generous support from many individuals and organizations. A grant from the TIAA-CREF Institute and a fellowship from the Center for Humanities and Arts at the University of Georgia enabled fieldwork and freed time from other obligations to write large sections of the book. Additional support, without which the work would not have been possible, was provided by the Institute of Higher Education, the Institute for Behavioral Research, the President's Venture Fund, the Office of the Provost, and the Department of Sociology at the University of Georgia. I thank William Finlay, head of the Department of Sociology, Paul Roman, center director at the Institute for Behavioral Research, and Steven Beach, director of the Institute, for particular creativity and resourcefulness in their support. I continue to marvel at Thomas Dyer, professor emeritus of history and former director of the Institute of Higher Education, for his ingenuity and grace. The welfare and work of a university are fostered and facilitated by people like him.

I presented portions of this work at the meeting of the International Sociological Association, Durban, South Africa; the annual meeting of the Association for the Study of Higher Education, Louisville, Kentucky; and at a conference on university faculty sponsored by the University of Georgia and organized by Libby Morris, director of the Institute of Higher Education, to whom I express appreciation for inclusion in the session. I acknowledge the especially helpful comments provided by Janet Lawrence and Jack Schuster at the event.

James Hearn and Sheila Slaughter, who have in common not only great minds but also unusual generosity, read the entire manuscript and provided numerous and invaluable comments and suggestions, for which I am certain the work is better off. Failings of it are mine. Linda Renzulli read and commented on the introduction, and she, in addition to Linda Grant, supplied alternatives for a title, pieces of which have been incorporated into the final rendition. There is no irony that my lunches and dinners with Barry Schwartz are permanently inscribed on my memory. One is able to make that observation of few people because few people possess the kind of originality and brilliance that he does. His quick insight, sharp wit, the unadorned pushing and unabashed prodding, and the unanticipated reminders, even of work close to me, such as the discussion of ideas developed by the sociologists Emile Durkheim and Robert Merton that appears in the final chapter, have made for long, bountiful, and priceless meals.

This book marks the second occasion on which I have had the pleasure to work with Douglas Mitchell, executive editor at the University of Chicago Press. The pleasure comes from the utter professionalism, which otherwise seems to decline by the day, but with which he, his assistant Timothy McGovern, and the review and production teams at the Press go about their work. I express very great thanks to the two anonymous reviewers who took the time to provide detailed comments and clear recommendations for the book.

Fieldwork often has its adventures, and there was no shortage of them as I traveled across the United States, by plane, train, and automobile, to talk to people about their work. There were the gas station–convenience store cashiers, hotel receptionists, train conductors, airline connecting-flight-information hostesses, lost-baggage-claim assistants, toll-booth clerks, parking garage attendants, rental car "hotline" operators, a tow-truck crew, late-night waiters, early-morning waitresses, and a highway patrol officer who helped me, for the most part, get to where I was going. En route to one of the field sites, in a numbingly cold month of January, I got stuck in Boston in a snowstorm of historic proportions. But by luck, my father-in-law, Eric Thorgerson, and his wife, Elizabeth Foote, lived in Cambridge. All the audio and recording equipment in tow, I trekked from the Logan Airport by T and then, in wingtips, by foot through the unplowed streets to their home for a stay without limit. For a warm dwelling, comfortable bed, food, the offer of new shoes, and, none the least, the company and conversation, thank you.

The greatest gift given to me by my father-in-law, however, has been his daughter. The world is made a little lighter, and much safer, by her. Remarkably kind and patient, my wife, Erika, is blessed with the discernible gifts of compassion, mercy, and forgiveness. I have been but one of the beneficiaries. There are many others: the world of living things unable to protect themselves, those trying to live, those having to die. She excused my days and weeks away, and the weekends and late nights with other preoccupations. She understood and tolerated my distractions, away and at home. She unreluctantly lived alongside the project day by day. She read, edited, and commented on successive drafts, despite severe demands on her time and even as she completed a book of her own. To the busyness of life, she brought a calm. Even when the days were at their most challenging, I always knew I could go home and find an environment that was renewing. For this and more: *Maximas ac plurimas gratias. Propter te, vita est quam optima.*

Looking back, my thoughts wandering over the years, I recognize debts of gratitude that run so deep their comprehension gives great pause. I think of and I thank my parents. Much about my life I owe to them. For almost twenty-five years, since leaving central Pennsylvania to become an undergraduate at the University of Chicago and subsequently thereafter, I have spoken with my father weekly and sometimes daily. It has been a gift I would wish for everyone. Much about it, though, is rare. And it has become clear to me, sometimes painfully so, that it is also irreplaceable. The awe of such a gift, however, resides not only in the realization that it has helped form me, but also that it will never leave me. It has been my fortune to find always an unfailing enthusiasm, a dogged encouragement, and an indefatigable resolve to move forward. For all the kindness, support, generosity, and companionship through the years, I am, and will forever be, sincerely grateful.

Shortly before I was to send in the manuscript for publication, my mother died. Her main career moved through a sequence of seven children. For fifty-five of her seventy-seven years she was a professional mother. The time alone turns out to be a reliable indication of moral commitment. Some of us know, but most can only imagine, that raising and running a household of seven children is not a job for the faint of heart. But her heart was strong, and its pulse was felt irrepressibly. Perhaps her greatest gift was an unusual ability to reach, understand, and inevitably sympathize with the interior lives of people. Her place in the world was predicated on a compassion for others and their human circumstances. She shared this gift with people of every age, but most especially, and with profound results, in the lives of the very young. Her affection, so abundantly given, brought assurance and, usually, a transformative perspective. My dedication of this book to her is but a very small expression of gratitude for much more that she gave to me. It stands as a permanent tribute to a life and to the labors of love.

Introduction

According to the United States Department of Labor, the person reading this sentence, if employed full-time in the American work force, will on average work 8.09 hours a day.[1] Extending this figure over time, one will work approximately 40.45 hours in a standard work week, excluding weekends, when a typical worker works an average of 5.80 additional hours.[2] In a given month, this individual will work roughly 175 hours, and in a given year, approximately 1,941.6, if one liberally subtracts four weeks for vacation, sick leave, or other absences.[3] Indeed, work's omnipresence has found its musings in literature, let alone the empirical sciences, as when William Faulkner was prompted to observe: "One of the saddest things is that the only thing a man can do for eight hours a day, day after day, is work. You can't eat eight hours a day nor drink for eight hours a day nor make love for eight hours—all you can do for eight hours is work."[4]

Faulkner appears to have been more energetic than the average American, for it is only hours of sleep that rival work hours, and not eating, drinking, socializing, playing sports, watching television, making telephone calls, paying bills, cutting the grass, or any of the other myriad activities on which the Department of Labor collects data. The typical full-time American worker will sleep, not on the job, an average of 8.26 hours a day, 41.3 hours per work week, 179 hours per month, or what amounts to 1,982.4 hours in the equivalent 48-week year.[5] Put differently, during waking hours, the average American, when not on vacation, sick, or cutting loose, will devote more than half of his or her time—51.4 percent—to work.[6] Around the world, exact working hours vary in the day, month, and year for the average worker, but from Korea to Iceland, Japan to Mexico, the hours are unambiguously substantial in magnitude.[7]

These figures mask important variations, however, lumping as they do workers of all kinds. Turning to specific sectors of work, one is apt to see significant differences in the number of hours workers devote to them, which reflect, among other things, the very centrality that work holds in people's lives. One such sector consists of the professions, such as law, medicine, and academe, the sector understood sociologically to contain those lines of expert work requiring protracted training and mastery of a technical body of theoretic knowledge.

The data bear this out. Weekly work among lawyers averages 49 hours[8], among physicians 50.8 hours[9], among academics 52.1 hours[10], each consuming an even larger share of that waking time available in a given week, month, or year. Yet these numerical estimates are, like all of the estimates above, but crude indexes of work and its larger significance to people's lives. Many people regularly work more, and others less, than these numbers tell, owing to numerous factors, such as type of employment and, particularly within professions, specialty area and form of practice. The numbers do not even necessarily record accurate measurement of work, the definition of work itself quickly blurring just as one begins to push the point. How does one factor, for instance, the time one thinks about work, and one's fellow workers, as well as the time it takes to get to and from work? The boundary between work and life outside it becomes increasingly blurred.

One point, though, is clear: work is a chief involvement of adulthood, and occupations are a major organizing feature of modern societies that structure and define much of any given individual's life course. We need to inquire, for specific theoretic reasons, about what all this work *means*—and about how such meanings may evolve over the course of a lifetime at work—to those in its grip, for indeed the grip lasts not just for a day or week or year but for decades, so as to compose a *career*. That is what this book will do.

This is a study of contemporary academic careers situated in varieties of the modern American university, as revealed in the lives of fifty-five university professors. I examine how members of one profession—academe—age in relation to their work, using a sample of scientists—physicists, specifically—who span a spectrum in age as well as types of institutions that have employed them.[11] For members of this sample, work is now measured by as many as four decades, and by no fewer than one. Some have worked in universities that demand much of them, and others in institutions that are comparatively less exacting. The spectra of age and institution allow us in this study to see what work means—and how these meanings have variously evolved—for people and their passages throughout an academic career.

Among studies of people and their work, including those of the academic profession, this one is unique: it is longitudinal. In 1994, I interviewed sixty physicists at universities across the United States. That study was published as *The Stars Are Not Enough: Scientists—Their Passions and Professions* (University of Chicago Press, 1998). This first work focused on scientists' ambitions, how their ambitions vary across academic institutions, and what academic institutions do to their ambitions over time in the course of a career. The study accounted for scientists' professional pasts and how they envisioned their professional futures to arrive at an evolving understanding of where and how they saw themselves, their careers, and their scientific contributions. The study thus established where scientists saw themselves professionally and where they saw themselves headed in their profession. Since this work forms the foundation on which the present follow-up study builds, I will refer to these works respectively as the *foundational* and *longitudinal* studies.

Ten years passed, and I completed another series of interviews with the same people. The present project is the first sociological study ever to follow the same professors and their careers over time.[12] The study approaches the lives of professors from a sociological framework: age and institutional location provide the structure to analyze individual, subjective careers through diachronic change. The study seeks to show how institutions organize careers and people's evaluations of their experience in these institutions.

A wealth of knowledge can be generated by a study that systematically tracks people to articulate the structure of their careers. Much remains to be discovered about how academics (and other professionals) experience work over a span of time, how they view their careers progressing (or failing to progress), and how institutional environments facilitate (or impede) such development. Longitudinal data add spatial and temporal dimensions to previous synchronic studies. Consequently, one is placed in a position of answering the following questions about academic careers, which motivate the present study:

1. What continuities and changes—in aspiration, satisfaction, motivation, commitment, and identification with work—mark the careers of scientists over a ten-year span of their careers? How do these continuities and changes vary by scientists' ages or career phases? How might these continuities and changes systematically emerge and pattern themselves by the type of institution in which scientists work?
2. What knowledge have these specific academics acquired about

themselves, their institutions, and the academic profession in ten years? Most important, how does the content of this acquired knowledge vary by individual age and type of university?

In attempting to answer these questions, this work seeks to provide a theoretic basis for evaluating the structure and experience of academic careers. As such, it will bring us closer to understanding how institutions shape careers. In broad terms, this book offers a rare, developmental view "inside" academe, its people, and the causes and courses of their careers.

GUIDING THEORETIC PERSPECTIVES

This work is situated at the intersection of three distinct bodies of work. So diverse are these bodies of work that they are rarely brought together, except, as in this case, when they may be used profitably to anchor and inform theoretically a subject that bears directly on each of them.

Given that the subject is about work, I turn to a perspective on *occupations* for theoretic guidance on people's outlooks and experience of work, focusing, as I will in this study, not simply on work hours but on what work means and what larger significance it might hold to the people performing it. Given that the subject is about how people age in relation to work, I turn to a perspective on the *life course* for theoretic guidance on how people's outlooks and experience of work evolve over time in an occupational career, examining, as I will, how such meanings of work change or remain the same. Finally, given that the subject is situated in a particular line of work—science in the academic profession—I turn to a perspective in the *sociology of science* for theoretic guidance on how careers take shape and unfold in this specific social-institutional setting.

The Occupational Perspective:
Careers, Identities, and Institutions

The sociological study of people and their work owes much of its origin to Everett Hughes, who is considered a father of the sociology of occupations (see Hughes 1958a, 1971, 1994).[13] From 1938 to 1961 at the University of Chicago, Hughes developed an approach to the study of work, closely connected to the Chicago School of Sociology, that located people and occupation ecologically, in their socially situated environments. This meant that to study people and their work, one went into the field, which in this period usually was the city of Chicago, the great laboratory of the Chicago School, in which to observe and study social process.[14]

Hughes's theoretic orientation to the study of work can be understood as interactionist. Occupation and individual, structure and self, institution and identity are created by the reciprocal interplay between macro and micro forces. This theoretic orientation specified an empirical method: anthropological-like field study involving heavy use of the interview and participant observation among the institutions and individuals who mutually create one another. The result tended to be a *biographical* form of data and analysis wherein the lives of individual workers came to life in the context of the institutional worlds in which they were enmeshed.

Hughes's influence can be measured in many ways, one of which is undeniably the vast legacy of students and their own work, sometimes begun as Chicago dissertations or master's theses, other times as early-career collaborations with Hughes, or subsequent independent forays. Theory and method combined typically in the case study (even as Hughes stressed the importance of comparative work on occupations). His students consequently produced a mosaic of work on occupations and occupational life: Blanche Geer on medical students (1961); Anselm Strauss on medical students (1961) and chemists (1962); Fred Davis on journalists (1951), Eliot Friedson on doctors (1960, 1970, 1975); Robert Habenstein on funeral directors (1954); Howard Becker on professional dance musicians (1949), Chicago public school teachers (1951), medical students (1961), and artists (1982); and, of course, Erving Goffman on psychiatric orderlies (1961), among many others.

A set of the earliest occupational studies in modern sociology laid the foundation for this work, conducted by first generation Chicago School sociologists, typically under the direction of Robert Park, Hughes's own doctoral mentor. From Anderson's study of hoboes (1923); Shaw's on jack rollers (1930); Cressey's on taxi dance hall girls (1932); Donovan's on salesladies (1929); Hayner's on waitresses (1936); and finally to Hughes himself on real estate agents (1928), one comes to see a colorful panorama of work ethnography across these intellectual generations. Indeed, Hughes's own professional goal was to inspire and himself conduct a massive array of occupational case studies, building a store of detailed knowledge on an ever-differentiating occupational system. He succeeded in inspiring a sociological generation that followed his lead. Even though extensive fieldwork was conducted and numerous case studies emerged, no comprehensive collection developed on the scale he once imagined. Nevertheless, an influential and enduring theoretic perspective crystallized around Hughes, a biographically based interactionist one that studied and understood work through the eyes and experiences of the people performing it (see, for example, the contemporary collection by Harper and Lawson 2003, in which sociologists study work and work-lives across an occupational spectrum).

In this perspective, the *career* exists as a core concept because it, among all concepts pertaining to work, best captures the element of time, and hence the process by which individuals and institutions are reciprocally constituted by interaction (Barley 1989; Van Maanen 1977). In one of Hughes's most notable formulations, careers are seen and studied for their "two sides" (Hughes 1937; later developed by Goffman 1961; see also Hughes's posthumous publication 1997). One side is the *objective career*, which consists of the sequence of statuses a person holds over time. The statuses may be indicated by positions or offices or titles: freshmen, sophomores, juniors, and seniors composing an educational career; second lieutenant, first lieutenant, captain, major, lieutenant colonel, colonel, and general composing a military career; assistant, associate, and full professor composing an academic career.

The second side, existing in tandem, is the *subjective career*, which consists of the shifting personal perspectives individuals develop about themselves and their work as the objective career unfolds (Stebbins 1970). How do undergraduates, for example, experience and understand each status that marks passages in an educational career? How are passages experienced by military officers, university professors, and all others whose lives and livelihoods are marked by movement from one status to another throughout biographical time?

Hughes marshaled the idea of *turning points* as a social mechanism that explains when and how change occurs in the subjective career as it engages in dialogue with the objective side (Hughes 1958b). As lives and careers transpire, people undergo a series of changes, not only in their objective status, but also in the patterned subjective views they hold about themselves in light of this change. The young assistant professor comes to see him or herself in a substantially new and different light from that understood as a student undergoing intense training and socialization for the professorial role, just as the emeritus professor—at the other end of a long sequence in professional status change—comes to see him or herself differently than viewed through the lens of a once-regular member of a senior faculty.

The idea can of course apply generally to status change, both objective and subjective, as when men and women progressively enter new statuses and consequently formulate new views of themselves because of them, moving for example, at the onset of adult life from single adulthood, courtship, marriage, parenthood, grandparenthood, to widowhood in life's final set of status transitions (Glaser and Strauss 1971).

For all the emphasis placed on process as a theoretic objective and on biography as a methodological tool to satisfy it, Hughes's work, and that which followed in its tradition, remained cross-sectional. In some respects, this is surprising, given the exceptionally heavy thrust on temporality as communicated by such concepts as the career and its mutually unfolding and interacting sides. In other respects, it is easier to understand, since the idea of longitudinal study in sociology would not pick up in earnest, or have many examples to illustrate its utility, until the 1960s, in the latter part of Hughes's professional work. But now coming at a time when longitudinal inquiry has matured, the present work will incorporate a longitudinal approach and attempt to see what strides can be made within an adjusted theoretic framework originating from Hughes.

This occupational perspective is relevant to the present concerns because it offers a way to understand and explain identity, or more specifically, *professional self-identity*, that is, how people know and understand themselves in light of their experience of work. Moreover, because the perspective is fundamentally temporal, adding the element of process to see how individuals are "made" (and perhaps sometimes "un-made") by institutions, one can also see how professional self-identities evolve as individuals experience turning points throughout the course of their careers.

The Life Course Perspective: Aging in Cohorts

The life course consists of the patterned progression of individual experience through time (Clausen 1986). Patterns in a progression arise from a "sequence of culturally defined age-graded roles and social transitions" that individuals enact from birth to death as continually socialized members of a society (Caspi, Elder, and Herbener 1990, 15). In life-course analysis, individuals are studied as they move along "pathways through the age-differentiated life span," where age differentiation refers to "expectations and options that impinge on decision processes and the course of events that give shape to life stages, transitions, and turning points" (Elder 1985, 17). In the realm of work, the life course may be applied to involve passage among roles in a career, loosely defined. I inquire about the ways in which such passage is structured, experienced, and understood by individuals and the larger work collectivities to which they belong.

The impetus for a sociology of the life course arose in large measure from prior, psychologically oriented attempts that conceived of development predominantly in terms of preprogrammed maturation, often promulgated in the form of stage theories of aging. Notable among the theories are Erikson's "eight stages of man," Levinson's "seasons of a man's and of a woman's life," and Sheehy's "passages" (Erikson 1950; Levinson 1978, 1996; Sheehy 1976). In their various formulations, developmental stages were asserted to be universal, inherent to human aging among men and women, from one society to another. Predecessors to these theories, again psychologically rooted, confined development to childhood: adulthood was merely an expression of the developmental scripting that had occurred or not occurred by the end of adolescence (for a review, see Mortimer and Simmons 1978).

For sociologists, however, aging should be seen and studied in contexts, those contexts not merely serving as a *setting* for but as a *constituent force* of development. Dale Dannefer's critical call to question the epistemological assumptions underlying stage theories of aging remains as key today as when it was proposed to much debate in the 1980s: "Why should a universal pattern have been a theoretically expected or desired claim to make in the first place? What mode of inquiry and what kinds of assumptions would lead one to assume such an invariance?" (Dannefer 1984a, 102–103; see also Baltes and Nesselroade 1984 and Dannefer 1984b). In this important theoretic sense, different contexts entail systematically distinct consequences for socialization and development throughout the entire life span. In the work arena, one can observe how behavior and beliefs understood as "developmental aging effects" may be, instead, a product of social structure—"an artifact of the organization of work settings and of norms about ideal careers" (Dannefer 1984, 110).

Cohort analysis has typically been used by sociologists and others to understand how individuals pass through time in socially patterned, yet distinct, ways. Cohort analysis seeks to avoid an "ontogenetic fallacy" of postulating universality in human development by investigating how groups of individuals age variously (Elder 1975). Glen Elder's work on the life course, for instance, locates individuals in historical times and socioeconomic contexts in order to see how development has transpired differently for cohorts proceeding through time under different environmental conditions. His studies of cohorts coming of age during and outside of the Great Depression illustrate the differential force of history and socioeconomic context on development in childhood and in subsequent adult phases (Elder 1974, 1981, 1998).

Cohorts may also be used, as they will in the present study, to investigate differential meanings of age, in particular, differential meanings of work and career as socialization and development occur. This approach, developed substantially through the work of Bernice Neugarten, emphasizes the normative underpinnings of age, those underpinnings forming an enduring component of culture (Neugarten 1968, 1979, 1996; Neugarten and Datan 1973; Neugarten, Moore, and Lowe 1965). In the words of Clair, Karp, and Yoels, this work inquires about "how persons occupying different locations in social space interpret and respond to repeated social messages about the meanings of age" (1993, vii). What work means to the senior physicist may be altogether different from the junior physicist or the physicist at mid-career—not only because these individuals occupy distinct age-graded statuses but also because they have likely experienced careers that have developed under substantially different institutional and sociohistorical conditions. What is more, if one takes seriously the idea that contexts shape development, then physicists (and other members of the academic profession) may see their careers, and their professional self-identities, develop in fundamentally different ways depending on their institutional location.

An age-graded life course is not tantamount to monolithic stages, proposed by most stage theories of aging discussed above, in which all individuals proceed through a largely invariant sequence of crises, challenges, or turning points. Rather, an age-graded life course presents the idea of a general conception and socially desired unfolding of lives through loosely defined periods of life. Some such periods characterize some individuals and not others. Individuals who enter and leave these periods do so at different rates. Some periods characteristic of a subset of people may be skipped altogether by another subset in light of differing sequences of events and turning points that socially situate their life passages. In short, the idea of a normative, age-graded life course more fully allows for the possibility of variation in the living of lives than do most stage theories of aging. And it is a goal of most age-graded studies of lives to examine how lives are variously, rather than monolithically, patterned and experienced. In order to emphasize the variability over the uniformity of lives and careers, and to avoid any confusion associated with the term *stage* and stage theories of aging, I will refer to *phases* when discussing the patterning of periods that characterize people's lives and careers. The idea of life phases seeks to underscore the wide-ranging permutations occurring in their enactment. These variations give rise to a more dynamic conception of how lives and careers are divergently patterned.

A crucial point is sometimes lost, even among those who embrace a life-course view responsive to the variations of social and historical time. Matilda White Riley has referred to it as the "life-course fallacy" (Riley, Foner, and Waring 1988). The life-course fallacy is a reminder that cohorts do not age alike: one cannot assume that what is typical for one cohort applies to other cohorts as they in turn age. This is so because conditions of the contexts in which people age are changing throughout time. Following a life-course view, this applies regardless of how one defines or operationalizes "contexts"—schools, families, marriages, occupations, and the like. A son or daughter, coming of age under his or her parents ten years after an older sibling, may experience and understand adolescence in a fundamentally different way from the older sibling, not just because of individual differences between the siblings but because of a changed matrix of relationships between parents and child, because of changed financial conditions that characterize a household as its breadwinners advance in their earning power, or for still some other temporally situated set of reasons that systematically differentiate one cohort of offspring from another.

Here, "contexts" denote the American university and the academic department. In the present case, following the fallacy, one would commit a conceptual and theoretic error if one were to postulate that patterns of self and career displayed, say, by senior physicists ten years ago would be those adopted by physicists now advancing into the same age-graded statuses and roles. The same would be true of the youngest scientists now advancing to middle phases of their careers, indeed to everyone passing into points of careers once occupied and experienced by a preceding cohort. In order to understand genuinely how academic (and all) careers take shape—and to explain age differences across cohorts—one must follow the subjects under study to build a stock of knowledge about the experience of aging in one's work.

This is done infrequently for reasons that are usually practical. Such research design requires some interval of time, typically many years if not decades, in order to study continuity and change in a specified set of behaviors or conditions. This can represent a substantial opportunity cost and therefore tends to run counter to the goals of most researchers. In addition, such research design can be prohibitively costly and difficult to fund, owing to the elaborate procedures, extensive infrastructure, and large research staffs often necessary to conduct the research. Finally, such research design is susceptible to subject attrition over time, which of course immediately compromises the goals of the research. For reasons explained in the next chapter, I encountered almost none of these problems in this study, but that may not be the norm in longitudinal work.[15]

This life-course perspective is relevant to the present concerns because it underscores the importance of cohort and context in situating the meanings of careers: it is here where meanings originate and acquire their social significance. These meanings may indeed be variegated given the developmental force that times and places can exert on the experience of unfolding careers. Theoretically, one is led by this perspective to consider how people *age within cohorts*, granting the opportunity to see how the meanings of aging in one's work might vary by the times of careers and the places in which they transpire.

The Sociology of Science Perspective: Stratification

If one attends to the forces that contexts can differentially exert on development in a career, then one must turn to the social-institutional arena that constitutes those contexts. In doing so, an important theoretic perspective is gained on how careers are specifically structured. In turning, therefore, to the sociology of science, I am drawn more particularly to a line of research concerned with work contexts, or strata, and their effects on scientific careers.

The heart of research on stratification in science is the *reward system* and its operation across the contexts of work in which science is done. In science, as in all social institutions, there is inequality. As Harriet Zuckerman has observed:

> Stratification is ubiquitous in science. Individuals, groups, laboratories, institutes, universities, journals, fields and specialties, theories and methods are incessantly ranked and sharply graded in prestige. Even rewards for assessed contributions are themselves graded. The topmost layer of each hierarchy is made up of an elite whose composition rests on socially assessed role performance or, in the case of fields, specialties, theories, and findings, on their cognitive standing. (Zuckerman 1988, 526)

In their various forms and magnitudes, rewards constitute recognition. Recognition in turn is centrally situated in the occupation of science as well as the lives and minds of perhaps all scientists, albeit in varying degrees. The centrality of recognition to the operation of science and scientists has been examined extensively, beginning with Robert K. Merton, whose celebrated work laid the foundation not only for this specific subject but for the entire sociology of science.

Merton explained that recognition is important to science—and scientists embark on quests for it—because recognition from those competent to judge a contribution is the prime indicator that a scientist has fulfilled the goals of science, to extend certified knowledge. "Recognition for originality becomes socially validated testimony that one has successfully lived up to the most exacting requirements of one's role as a scientist" (Merton 1973a, 293). Recognition is thus institutionalized: it is both essential to progress and therefore expected in trained individuals, if as socialized members of this profession they seek to satisfy the goals of science.

Cumulative advantage and *cumulative disadvantage* is a theory developed by Merton and elaborated by others to explain inequality in science. At root, the theory explains how increasing disparities come to characterize the "haves" and "have nots" over the course of a career in science (and conceivably in other institutional domains).

> Processes of individual self-selection and institutional social-selection interact to affect successive probabilities of access to the opportunity-structure in a given field. . . . When the role-performance of an individual measures up to demanding . . . standards . . . this initiates a process of cumulative advantage in which the individual acquires successively enlarged opportunities to advance his work (and the rewards that go with it) . . . [those who find their] way into [elite] institutions ha[ve] the heightened potential of acquiring differentially accumulating advantage. (Merton 1977, 89; quoted in Zuckerman 1988, 531)

In short, the theory holds that "certain individuals and groups repeatedly receive resources and rewards that enrich recipients at an accelerated rate and conversely impoverish (relatively) the non-recipients" (Zuckerman 1977, 59–60). In still plainer words, the rich get richer at a rate that makes the relatively poor become even poorer (for an extended consideration of the theory, see Zuckerman 1998; DiPrete and Eirich 2006).

The "Matthew Effect" elaborated by Merton is a special case of cumulative advantage. Named after the Gospel of St. Matthew, it holds that already-recognized scientists receive disproportionate recognition for subsequent contributions. "Eminent scientists get disproportionately great credit for their contributions to science while relatively unknown scientists tend to get disproportionately little credit for comparable contributions" (Merton 1973b, 443), or, following the Gospel, "For unto every one that hath shall be given, and he shall have abundance; but from him that hath not shall be taken away even that which he hath" (Matthew 25:29).

According to the broader theory, early access to resources is key. "Early" is a necessarily unspecified point in time, since resources available to some sooner than others—in graduate training, college, or even pre-collegiate experiences and conditions—can be used to begin the spiral of advantage on the one hand or, conversely, disadvantage on the other. The point is that advantages in early career, and in early-life phases, position individuals for further achievement, which in turn brings its own rewards and resources that can be put to use for still further achievement, thereby establishing a process that over time significantly differentiates careers, their development, and conceivably the professional self-identities of those experiencing these careers. Thus, recalling Hughes, not only are different careers produced, but presumably different people as well. It is an empirical question of what becomes of such scientists' commitments, motivations, and identifications, not to mention their definitions of and expectations for success, as they witness their variously rewarded and recognized careers unfolding alongside this social process.

The stratification perspective in the sociology of science contains some assumptions, among them the following two:

First, it seems that all, or most, members of science are equally and evenly socialized to pursue recognition, that is, to pursue research careers devoted to scientific discovery. By this view, comparative successes and failures over the course of a career can be explained by accumulative (dis)advantage, situated in a reward system that distributes recognition on preponderantly universalistic criteria (e.g., Allison, Long, and Krauze 1982; Allison and Stewart 1974; Cole and Cole 1973; Gaston 1978; Long 1978).

Second, and flowing from the first point, the perspective evaluates role performance on the basis of a singular system of reward, that oriented to research. Moreover, this reward system, geared as it is around the extension of socially certified knowledge, seems to refer to a specific subset of scientists—*elites*—or those habitually engaged in scientific research over extensive portions of their, if not their entire, career. Yet reward systems, even in research universities, are plural and encompass roles and performances other than research. This may be more true in some types of universities than in others, but even in the research sector of higher education, one can observe this variety (Blackburn and Lawrence 1995; Clark 1987; Finkelstein, Seal, and Schuster 1998).

The stratification perspective may contain ways to explain this phenomenon. For example, the existence of competing reward systems, such as in teaching, service activities, and administration, may be explained as an institutional mechanism that allows organizations to accommodate failure in a "focal role" of research (cf. Glaser 1964a; Goode 1967). But such an explanation would need to account for how and why this occurs, particularly in institutions—research universities included—that espouse multiple organizational missions. I shall take up this problem in the body of the present work.

It may also be the case that individuals identify more strongly with non-research roles, in the instances where this may be observed, for reasons having little or nothing to do with accumulated (dis)advantages in a research career. For many academics, research careers may hardly exist, with few of them engaged in steady publication, as Bernard Gustin argued (Gustin 1973). In still other instances, research careers may be embraced strongly, perhaps even within scientific strata where the feedback of rewards and resources is comparatively meager and where therefore the theory of cumulative (dis)advantage would predict such careers would drop off. Furthermore, these identifications may change, sometimes in substantial ways, over the course of a career (Glaser 1964b; Pelz and Andrews 1966). In the absence of tracking scientists over time, and obtaining data on how they perceive and account for their careers, one is at a loss in understanding these career dynamics. These questions, and the larger theoretic issues they address, remain to be subjected to empirical inquiry.

This sociology of science perspective is relevant to the present concerns because it provides a framework for asking and inquiring about "what life is like" across strata, or contexts, of science, and what work comes to mean to people whose careers are situated in these strata over many decades of professional events and experiences. Harriet Zuckerman has noted this need in structural terms, though to date it has remained largely unexplored territory on all counts: "An important next step for research . . . requires more intensive study of the *organizational* bases of stratification in science—the ways in which processes internal to organizations work to rank and reward scientists" (Zuckerman 1988, 532–533, emphasis added). Thus, overall, this perspective highlights a need to investigate strata of science more fully, perhaps seeing ways in which differentiated structures and cultures of work arise within them, with varied systems of reward and correspondingly varied individual career orientations that may change over time.[16]

Why should one be concerned with careers in science or in higher education more broadly? The answer, while clear, is anything but plain: the advancement of science and of knowledge more broadly is directly contingent on those careers. It is only because of what individuals do in the course of their careers, situated as the reader shall see in institutions that foster and facilitate them in varied ways, that progress in any social-institutional domain can be made. What becomes of law and of people's legal fates, or of medicine and people's medical welfare, in light of conditions of careers that favor or stultify progress in these domains? What becomes of clients and patients unable to be served by learned and capable practitioners? The same may be asked of science, the future of knowledge and its sweep of beneficiaries. If culture and civilization are to advance, and if its citizens are to benefit in all kinds of social, cognitive, and economic ways, they must rely on both institutional and individual development in science.

If institutional contexts of science facilitate or, conversely, impede scientific (and/or other forms of academic) work more so than others, this is important data to bring to bear on the conditions of scientific progress or the lack thereof. And even though one might observe the research role as less central to the matrix of reward systems in some institutional contexts or in some phases of the career than in others, it remains a paramount role on which all other academic roles are contingent, for reasons that are also clear. "The research role," as Merton and Zuckerman have elegantly stated, ". . . is central, with others being functionally ancillary to it. For plainly, if there were no scientific investigation, there would be no new knowledge to be transmitted through the teaching role, no need to allocate resources for investigation, no research organization to administer, and no new flow of knowledge for gatekeepers to regulate. . . . The heroes of science are acclaimed in their capacity as scientific investigators, seldom as teachers, administrators or referees and editors" (Merton and Zuckerman 1973, 520).

To study individual careers, then, harking back to Hughes, is to study something larger than individual fates: it is to study institutions and the process of their creation and sustenance over time. Individual fates are most certainly crucial, and to be sure the reader will become well acquainted with them, for the larger institutional workings depend on their success, failure, or some intermediate fate. In science, as in all social arenas, one may view institutions and glean their paths through the eyes and experiences of individuals and their careers.

PHYSICS AND PHYSICISTS

Among professions, academia is uniquely situated in society as the profession that trains people for all other professions, and numerous other lines of work requiring certified education. In their roles, academics may thus be viewed as guardians of culture: they uphold cognitive and behavioral standards that have been created by their professional-disciplinary communities to ensure competent role performance (Parsons and Platt 1973). In addition, they guard culture by upholding a set of generalized ideals (Ben-David 1972). As masters in their various roles, they seek precision and excellence and, in principle, to inculcate these characteristics in their student-clientele so as to produce higher learning and a more advanced civilization.

To this end, professors assume a privileged place in the social organization of modern societies. Theirs is a profession on which society depends greatly, not only because all other professions and other lines of work are contingent on it, but also because it more generally and pervasively transmits a higher learning to all. This includes not only those who, at various points in their lives, enroll in the higher education system and whose intellectual and occupational futures are shaped by this learning, but also those throughout society whose lives are conditioned by a social-institutional order that is influenced and partly determined by education (Ben-David 1963, [1968] 1991). To examine professors' experience of and outlooks on their roles over the course of a career, then, is in effect to study the health and functioning of a centrally situated profession, from the vantage point of those chiefly involved in the performance of its craft.

Within the academic profession, the field of physics has its own situated significance, which draws me to account for why I have chosen it as the base from which to study academic careers. I originally chose to study physicists for specifically theoretic reasons. Physics is the oldest and most mathematical of the sciences, and thus is commonly considered the scientific discipline *par excellence*.

Physics, I put forth, stands between the sacred and the secular: physicists seek verifiable answers about a limitless universe, as if to bring ordinary mortals closer to the sublime and incomprehensible powers of the divine (cf. Paul 1980). Possible answers about the universe in far-off space and time bring us, akin to spirituality, closer to the extraordinary powers of the unknown. It is for these reasons that physics—the field of formulas and abstruse numbers—is said even to possess romance and beauty (Chandrasekhar 1987; Glashow 1991; McCormmach 1982). Most important, physics has a recognizable genealogy of near-immortals—Ptolemy, Copernicus, Kepler, Newton, Einstein—who promote a kind of heroism both by serving as models of great attainment and providing the contours of a paradigmatic academic career. If one is interested in how careers take shape and unfold, physics provides an ideal setting to study how this occurs variously in the midst of such strong cultural conceptions of what constitutes a superlative career.

If one harbors any doubt about the mythic quality of physics and a physics career, a pilgrimage to the local bookstore furnishes an ample form of cultural evidence in support of these claims. Nowhere on the shelves of a standard store that sells books does one find the kind devoted to physics and to physicists. No other academic discipline or field shares the mythic notoriety of physics and its place in the public mind. The evidence presented in table 1 provides a public form of testimony to the "myth of physics and physicists." In it are listed a sampling of book titles, allotted to the subject of physics, obtained on a random Saturday afternoon while looking at the shelves of a franchise bookstore.

Examining the titles, I note several patterns: a connection between physics and religion and/or spirituality; the idea of a search for deep answers about fundamental life questions; the notion that the answers are measurable through rational design; an anticipation that such answers would bring an "order" as yet unknown and unrealized in a modern and fragmented world; a recognition that such pursuit, and those engaged in it, brush with "greatness," as if to be god-like; and, finally, the ultimate aim, or perhaps hope, that such revelation might entail a kind of immortality—or, at minimum, that the search may yield an ultimate permanence of understanding, of oneself, the world, and one's place in it, all based on *science*. This forms a broader line of thought in Iwan Rhys Morus's (2005) historical account of disciplinary imperialism, *When Physics Became King*.

Searching the shelves in sociology, anthropology, business, literature, music, art, biology, genetics, and chemistry, the titles are altogether different, devoted to specific subjects and strategies of the ordinary world. Indeed, in modern bookstores, a kind of cultural mirror on society's place and time, the only shelves that approximate titles found in physics are those found on shelves devoted to religion, philosophy, and psychological "self-help." The difference, though, is key: it is a difference between perpetual

TABLE 1. Sample titles and authors of popular books in physics

Title	Author
The Fabric of the Cosmos: Space, Time, and the Texture of Reality	Brian Greene
The Elegant Universe: Superstrings, Hidden Dimensions, and Quest for the Ultimate Theory	Brian Greene
Rational Mysticism: Spirituality Meets Science in the Search for Enlightenment	John Horgan
Stephen Hawking: Quest for a Theory of Everything-The Story of His Life and Work	Kitty Ferguson
Einstein and Religion	Max Jammer
The Fabric of Reality: A Leading Scientist Interweaves Evolution, Theoretical Physics, and Computer Science to Offer a New Understanding of Reality	David Deutsch
Measuring Eternity: The Search for the Beginning of Time	Martin Gorst
Fire in the Mind: Science, Faith, and the Search for Order	George Johnson
God and the New Physics	Paul Davies
Great Physicists: The Life and Times of Leading Physicists From Galileo to Hawking	William Cropper
The Scientists: A History of Science Told Through the Lives of Its Greatest Inventors	John Gribben
On the Shoulders of Giants: The Great Works of Physics and Astronomy	Stephen Hawking
The Road to Reality: A Complete Guide to the Laws of the Universe	Roger Penrose
Superstrings and the Search for the Theory of Everything	E. David Peat
Quantum Questions	Ken Wilber
Einstein and Buddha	Thomas J. McFarlane
The God Particle: If the Universe Is the Answer, What Is the Question?	Leon Lederman
The Physics of Consciousness: The Quantum Mind and the Meaning of Life	Evan Harris Walker
Parallel Worlds: A Journey Through Creation, Higher Dimensions, and the Future of the Cosmos	Michio Kaku
Doubt and Certainty: The Celebrated Academy, Debates on Science, Mysticism, Reality in General on the Knowledge and Unknowable, with Particular Forays into Such Esoteric Matters as the Mind Fluid, the Behavior of the Stock Market, and the Disposition of a Quantum Mechanical Sphinx, to Name a Few	Tony Rothman & George Sudarshan
Quantum Physics and Theology: An Unexpected Kinship	John Polkinghorne

belief and reality. Physics is ascribed to be a field that can, unlike any other, provide answers to complex metaphysical questions by employing *empirical* methods. Answers are knowable and verifiably true. Religion, philosophy, and psychological self-help all possess weaker and often nonexistent empirical claims.

Who can recount the heroes who populate the pantheons of sociology, anthropology, biology, chemistry, education, or art? All these and other fields have their respective pantheons, to be certain. But they are known by people idiosyncratically. For most others outside these fields, they are hardly known, if known at all. And which fields are the exceptions? The Platos, Aristotles, Kants, and Nietzsches of philosophy. The Buddhas and Confuciuses of religion. The Freuds and Jungs of psychology. Empirically, however, these pantheons are built on a ground less firm than physics. The larger point is that physics has a privileged place within academe, as academe occupies a privileged place among professions. Physics thereby becomes a noteworthy field in which to examine the unfolding of careers and how they are experienced by those working among such mythic company.

ORGANIZATION OF THE BOOK

The remainder of the book will trace the unfolding of careers in science and, in so doing, examine what scientists learn about themselves, their careers, and their profession over a significant interval of time. In chapter 1, "Following the Scientists," I will explain the research methods and design used for the longitudinal study. In addition, I will explain the rationale for the ten-year interval to study academic careers. Finally, I will characterize and socially situate the cohorts of scientists and institutional contexts of science studied by calling forth quantitative data on these dimensions.

The body of the book proceeds to examine aging in relation to work by investigating the patterns of careers that characterize each of three cohorts of scientists, situated in their respective institutions. Thus, in chapter 2, "Early- to Mid-Career Passages," I will examine the unfolding careers and attendant events and experiences that socially situate the youngest cohort of scientists in three prototypes of the American university. In chapter 3, "Mid- to Late-Career Passages," I will do the same for the middle cohort of scientists who work within the spectrum of institutional types. In chapter 4, "Late- to Post-Career Passages," I will do likewise for the eldest cohort of scientists, which will include discussion of how careers in science end and the ways in which scientists make transitions out of them into retirement, how these transitions are experienced, and what they mean to individuals.

In the final chapter, "Lives of Learning," I will provide an overview of results of the study, exploring what has been learned about careers in science, based on what scientists have learned from their own unfolding careers. I will discuss a theoretic framework that brings together each of the cohorts and each of the contexts of work in order to conceptualize the system of academic careers. I will also discuss how the results of this study may pertain to careers in other fields within the academic profession. I will conclude by examining what the results of the study suggest about the future of careers in science and the institutional settings in which academic science is done.

CHAPTER ONE
Following the Scientists

In this chapter, I explain the research design of the study. I describe the sample that provided the data to analyze and that forms the basis of the research. I tell of the methods I used and of the procedures I followed to learn about academic careers, how people experience them, and how this experience is structured by institutional contexts and age. I also describe the institutions themselves and account for ways in which they have changed in the ten-year period since I first began interviewing people. In short, I trace for the reader the steps I took to produce the present work.

Fieldwork for the foundational study on which I base the longitudinal work commenced and concluded in 1994. The book reporting the bulk of the first study's results was published in 1998. Fieldwork for the longitudinal study commenced in May 2004 and concluded in February 2005. The writing of the present book began immediately thereafter. The time between the first and second points of data collection thus spans ten years. How did the project get started?

FOUNDATIONS OF THE FOLLOW-UP STUDY

This book builds directly upon my earlier work, *The Stars Are Not Enough: Scientists—Their Passions and Professions* (1998), where I first examined careers in science using institutions and cohorts to investigate variation in the experience of academic work. In 1994, I designed a fieldwork project in which I studied the careers of scientists as situated in a variety of universities and career phases that spanned early to middle to late career. *Time*—as indicated by career phase—and *place*—as indicated by university type—thereby were theoretic anchors to investigate individual experience in careers.

To that end, these anchors sought to address, in ways responsive to the guiding theoretic perspectives discussed in the introduction, how the structure and experience of careers might be permutated by the specific types of institutions in which individuals work and by their professional age. Distinct work contexts and cohorts formed the design of the research in order to avoid portraying a scientific career as simply a monolithic series of stages, characteristic of stage theories of aging as previously discussed, in which individuals pass through a uniform set of preprogrammed points through the life span. Instead, the research sought to demonstrate how distinct contexts of academic work serve differentially as forces of, and not simply as settings for, development and how these forces are variously, as opposed to uniformly, expressed across phases of careers.

Six universities across the United States composed the original sampling frame, from which individual professors were selected randomly from respective university physics departments. Departments themselves were selected on the basis of their ranking in the assessment of graduate programs conducted by the National Research Council (NRC) (Jones, Lindzey, and Coggeshall 1982; Goldberger, Maher, and Flattau 1995).[1] I selected departments of physics that ranked at or near the top, middle, and bottom in the NRC assessment. I did so in order to establish the widest parameter of variation in careers. That is, top, middle, and bottom departments were built into the study design to permit comparison and contrast of scientific careers that are experienced under different structural and cultural conditions—the prevailing resources and expectations that situate and help define each of these types of departments and corresponding universities.

In turn, for comparative analytic purposes, the six universities were codified into three prototypes of institutions in order to elucidate the prevailing norms that structure academic life in what amount to distinct *academic worlds*. The three prototypical institutions, construed sociologically as academic worlds of differing beliefs and behavior about normative careers, are: the *elite world*, the *pluralist world*, and the *communitarian world* (for the original elaboration of these worlds, see Hermanowicz 1998, esp. chapter 2; 2005). These collective identities were *derived from*, as opposed to introduced before, the data analyzed in the foundational work. They thus are *based on* the data and have served as a way in which to organize the presentation and analysis of research findings. This is the reason I present them at the outset of this study.

In all three of these worlds, as in nearly all four-year higher education institutions, one finds the institutionalized triumvirate of roles so pervasive as if to become a cliché: research, teaching, and service. This is fully expected, since all of the institutions form parts of a socially regularized system of higher education. What is important for the present concern is not similarities, but differences, for it is in differences that contextual effects may be exerted on and observed in the individuals who "inhabit" these structures and cultures of work. Thus I call attention to the *premiums* the respective worlds assign to these roles, most particularly to research, because this role is least constant (and therefore most different) among the institutional types.

The elite academic world consists of those institutions that place the highest premium on research. Typically, the elite academic world consists of the private research universities and some prestigious public research universities. The overriding organizational goal is to garner additional prestige through the research and scholarly achievements of a faculty. Such universities often go to lengths to attract, retain, and compete among each other for "stars," those academics who have acquired wide notoriety. Most, if not all, departments in such institutions run doctoral programs, and their students are recruited from a nationally and internationally competitive pool of candidates. Examples of elite institutions include Caltech, Harvard, Princeton, the University of Michigan, and the University of California–Berkeley.

The pluralist world consists of those institutions that place a premium on both research and teaching. Typically, the pluralist academic world consists of the public research universities. The pursuit of additional prestige through faculty research and scholarly achievement is a goal of this type of institution, but not an overriding one. Such institutions typically employ a faculty that is more variegated in its goals and in its achievements. While some highly accomplished academics and sometimes even "stars" work in this type of university, they are a decided minority. This type of university normally does not compete to hire "stars," first because they are out-competed by elite institutions and because they may lack the resources and an organizational priority to recruit such individuals. Most departments in this type of institution confer master's and doctoral degrees, and tend to recruit students from a mixed regional-national pool of candidates. This type of institution answers to more varied demands, including mass teaching and service to communities and states. Examples of pluralist institutions include the University of Maryland, the University of Kansas, and Purdue University.

The communitarian world consists of those institutions that place a premium on teaching in the presence of research. Faculty in this type of institution often engage in research or scholarship of some kind and to some degree, at least at some point in their careers. In instances, highly accomplished researchers and scholars work in this type of institution. Teaching, however, is the overriding organizational goal, and an allegiance to that

TABLE 2. Interview questions on career success and failure

I. GENERALIZED DEFINITIONS OF SUCCESS LADDERS

1. What do you associate with a "successful" career in physics?
2. What do you think are the most important qualities needed to be successful at the type of work you do?
3. What does *ultimate* success mean to people working here?
4. Is there an understanding of a *minimum* needed in order to maintain respect among people here?
5. Is there an understanding of a *failed* career among colleagues here?
6. Taking your colleagues in this department, how would you say their success varies? Probe: Have they advanced at the same rate?
7. Where do you place yourself among that variety?

II. CONCEPTIONS OF FUTURE AND IMMORTALIZED SELVES

1. What do you dream about in terms of your career?
2. What ultimate thing would you like to achieve?
3. How do you envision yourself at the end of your career?
4. How would you like to be remembered by your colleagues?
5. What about your life do you think will outlive you?

Source: Hermanowicz, Joseph C. 1998. *The Stars Are Not Enough: Scientists—Their Passions and Professions.* Chicago: University of Chicago Press, 211–213.

goal defines the overall collective identity—loyalty to local concerns. Some departments in this type of institution may possess master's programs, and others may possess doctoral programs in addition, but many other departments confer only undergraduate degrees. This type of institution typically recruits students from a local or regional base, including some of the students in its graduate programs who, upon earning their degrees, may remain and gain employment in the region. Typically, the communitarian academic world consists of more regional public universities. Examples of communitarian institutions include the University of Tulsa, the University of Louisville, and Wichita State University.

Categorization into the types was based on individuals' responses to a subset of twelve questions composing an interview protocol used in the original study. These questions dealt specifically with career success and failure. They are presented in table 2. The full interview protocol used in the foundational study is presented in appendix A. Reviewing the twelve questions, one is able to see that, while these are posed specifically to physicists, the questions are generic. Since career success and failure are generally germane concerns and indeed institutionalized as such by organizational reward systems, they are fitting for a wide range of academic and professional fields.

I asked individuals to discuss conceptions of career success and failure in order to draw characterizations about what each means within their respective departmental and institutional contexts. Academic worlds cannot exist in the absence of a system of norms. Norms found in academic worlds govern individual behavior and enable group members to render judgments about career performance (Geertz 1973; 1983). Durkheim ([1915] 1965) used the term *moral order* to refer to the social arrangement of valued actions, beliefs, and orientations held by members of a group. Moral orders convey how life (or a career) within a group ought to be lived. For the study of academic careers, moral orders inform us about how careers are structured, interpreted, and understood among differing institutional types. In the case of academics, moral orders may be gleaned by how members of varying groups define and verbalize achievement, extraordinary achievement, and lack of achievement.

Elite, pluralist, and communitarian academic worlds are arrayed on an institutional continuum, as illustrated in figure 1 (Hermanowicz 1998; 2005). Elite institutions are at one end, communitarian institutions at the other, and pluralist institutions in between. The elite world occupies the space on the academic continuum where research is given particular emphasis in the presence of teaching. The communitarian world occupies the space on the academic continuum where teaching is given particular emphasis in the presence of research. The pluralist world occupies the space on the academic continuum where research and teaching are *either* given roughly coequal emphases *or* where large fractions of individuals alternately stress one over the other, so as to create a collective hybrid. This continuum of academic types thus reflects sociocultural distances and differences in the roles that departments and institutions socially certify as valid.

As the idea of a continuum conveys, there is overlap among the academic worlds. The overlap exists in part because of the similar baseline roles and missions among institutions and in part thereby because some subset of careers found in one world are similar to a subset of those found in another. Returning to figure 1, one thus observes, for instance, that some subset of communitarians may resemble a subset of elites, and vice versa. Such types can be identified as "communitarian elites" and "elite communitarians." A subset of pluralists may resemble subsets of elites — "elite pluralists" — or

FIGURE 1. Continuum of academic worlds

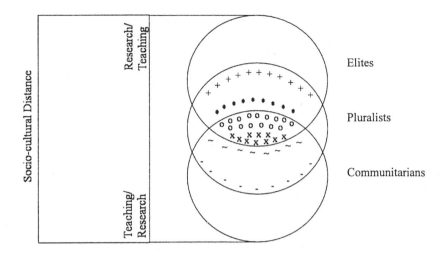

o = Elite communitarians
x = Communitarian elites
+ = Elite pluralists
- = Communitarian pluralists
~ = Pluralist communitarians
* = Pluralist elites

communitarians, "communitarian pluralists." Or, as their collective identity testifies, pluralists may represent pluralistic hybrids of the two polar forms—"pure pluralists." A subset of elites, less oriented and successful at research, may resemble of subset of pluralists, "pluralist elites."

In the sample, departments from two institutions comprise the elite world. One of these institutions is private, the other public. A department from one institution, a public university, comprises the pluralist world. Departments from three institutions, all public, comprise the communitarian world. The specific number of schools of each type used was a manifestation of obtaining a roughly equal number of respondents across the types. Thus, because communitarian departments tend to be smaller than either elite or pluralist departments, more were contained in the sample to bring about roughly equivalent subsamples of individuals across institutional types.

It is important to note that figure 1 illustrates an arrangement of several units. Let us be clear that the primary unit of analysis in the present research is the individual scientist and his or her career; the research herein asks how institutions, including the departments within them, affect careers.

The continuum illustrated in figure 1 depicts an arrangement not only of careers but also of departments and institutions. In the cases of all three units, one is able to say that careers, departments, and institutions tend to be more elite, pluralist, communitarian, or some hybrid thereof.

There is not always consistency of identity among the units. That is, one can think of elite careers of individuals who work in communitarian or pluralist departments or institutions. One can also think of communitarian careers of individuals who manage to be employed in an elite department or institution. Likewise, one can think of communitarian or pluralist departments in what are otherwise elite institutions. These instances may be relatively uncommon, but they do exist empirically.

The classification scheme observes a logic wherein *institutional* identity may be derived from the preponderance of its *departmental* identities. Thus institutions may be called elite if a preponderance of their departments are also elite. Institutions may be called pluralist if a preponderance of their departments are pluralist, that is, departments that stress a research-teaching hybrid or departments that are either elite or communitarian but whose collective result is plurality. Institutions may be called communitarian if a preponderance of their departments are communitarian (see Hermanowicz 2005).

Correspondingly, departmental identity may be derived from the preponderance of individual careers, which reflect levels of achievement. Departments may be called elite if a preponderance of the individual careers contained in them are also elite, for ultimately it is the records and achievements of individuals' careers who collectively determine a higher, aggregate identity. Departments may be called pluralist if a preponderance of individual careers are pluralist, that is, careers that stress a research-teaching hybrid, or careers that have stressed one or the other at different times whose result is a plurality of focus and orientation. Departments may be called communitarian if a preponderance of individual careers are communitarian, that is, having a preponderant orientation to teaching and other local service roles.

Data from the NRC assessment of doctoral programs explicates further the link I draw between departments and institutions. A total of six physics departments were used to sample physicists for the present study as stated above. Four of these departments were assessed by the NRC because they met NRC assessment criteria: they each contained a doctoral program with at least ten graduates within a specified time period. The remaining two departments, both communitarian, were included in the present study because they closely resembled the one other communitarian department that was assessed by the NRC. One of these two departments had a doctoral program but fewer than ten graduates in the specified time period, and so was not assessed. The other communitarian department had a doctoral program but discontinued it prior to the NRC assessment. It continued to maintain a master's program. These two departments were added to the present study to supplement the other communitarian department such that a sufficient number of scientists could be sampled from this departmental type. As I explained in the foundational work, the two communitarian departments not assessed by the NRC would in all probability score closely to the one that was, given the general similarity in program size and other departmental characteristics (Hermanowicz 1998, 226–227n3).

For each of the four institutions from which I selected the assessed physics departments, I obtained the rankings in *all of their other* programs assessed by the NRC. The total number of programs that the NRC assessed at this time was thirty-two. Not all of the four institutions maintained doctoral programs in all of these areas, and in some cases an institution maintained no such program/department whatsoever (e.g., the University of Houston, while it has a doctoral program in physics, has no classics department and thus no ranking for classics).

Taking the scores of all the assessed programs in the four institutions, I ranked them into thirds: top, middle, and bottom. Table 3 presents the results of the correspondence between departmental identity and institutional identity.

In what I call an elite institution, all seventeen of one of the school's assessed departments (or programs) ranked in the top. In the other elite school, a vast majority (22), or 73 percent, of its assessed departments ranked in the top. In what I call a pluralist institution, a vast majority of its assessed departments (28), or 87.5 percent, ranked in the middle. Four, or 12.5 percent, of its assessed departments ranked in the top. In what I call a communitarian institution, a vast majority of its assessed departments (6), or 85.7 percent, ranked in the bottom.

Thus, empirically, there is a general correspondence between institutional and departmental identity. Elite institutions are those with *mainly* top-ranked, or elite, departments. Pluralist institutions are those with *mainly* middle-ranked, or pluralist, departments. Communitarian institutions are those with *mainly* bottom-ranked, or communitarian, departments.

At the time of the original fieldwork, individuals were sampled from departments to cover early, middle, and late phases of careers and correspond-

TABLE 3. Correspondence between departmental identity and institutional identity, by National Research Council rankings

Institution	Number of Assessed Departments	Bottom Ranked	Middle Ranked	Top Ranked
Elite School 1	17	0	0	17
Elite School 2	30	0	8	22
Pluralist	32	0	28	4
Communitarian	7	6	1	0

Note: Based on the assessment of research-doctorate programs in the United States conducted by the National Research Council and reported in Jones, Lindzey, and Coggeshall (1982). Jones, Lindzey, and Coggeshall evaluated programs (departments) according to six criteria: program size; characteristics of graduates; university library size; amount of research support; scholarly reputations of faculty and programs; and records of publications attributable to departments. As in the foundational study (Hermanowicz 1998, 30), I took one indicator—scholarly reputability—to rank departments. "Scholarly reputability" represents the most general statement about departments, and thus is the most inclusive of measures, that was used in the assessment. Moreover, had the basis of rankings been predicated on any of the other five criteria or on all of them, the outcome would be no different: each of the six quality measures are highly intercorrelated (see Jones, Lindsey, and Coggeshall 1982, 166, table 9.3). Scoring on the scholarly reputability measure was between 0 and 5. Thirds were constructed using the scores. Thus, "bottom ranked" corresponds to departments/programs with a score between 0 and 1.6. "Middle ranked" corresponds to departments/programs with a score between 1.7 and 3.3. "Top ranked" corresponds to departments/programs with a score between 3.4 and 5.0.

ingly distinct professional ages. Career phases were a design of the research in order to emphasize variability in experience and structure of the career, again striving to encompass variation over uniformity in the characterization of people's experience of work.

Career phase/professional age was operationalized in the foundational study by the year in which scientists received their PhD to form three cohorts: scientists who received their PhD prior to 1970, thereby comprising those in late phases of the career at the time of the original research; scientists who received their PhD between 1970 and 1980, thereby comprising those in middle phases of the career at the time of the original research; and scientists who received their PhD after 1980, thereby comprising those in early phases of the career at the time of the original research.

A total of sixty physicists were interviewed in the foundational study between June and November 1994. Sixty physicists were interviewed because,

TABLE 4. Number of scientists in foundational study, by type of academic institution and cohort

Institution	Cohort (by year of PhD)			
	Pre-1970	1970–1980	Post-1980	Total
Elite				
School 1	8	6	6	20
School 2	1	—	2	3
Pluralist				
School 1	6	5	7	18
Communitarian				
School 1	3	3	3	9
School 2	3	2	2	7
School 3	1	—	2	3
Total	22	16	22	60

by this number, a "saturation point" in the data had been achieved (Morse 1994; Rubin and Rubin 1995). That is, data began to repeat themselves in predictable fashion, thereby rendering additional cases unnecessary to the goals of the research. By this measure, additional cases would add only marginal empirical and theoretic value to the study, yet entail significant costs in time and money. The response rate in the foundational study was 70 percent.

Moreover, the sample of physicists was arrayed roughly evenly across the three academic worlds and across the three cohorts; on the order of about twenty individuals for each of the three academic worlds and for each of the three cohorts. The foregoing explanation of the foundational study's research design is illustrated in table 4.

Ambition, defined as a strong desire to achieve, formed the theoretic focus of the foundational work. Since recognition is central to academic careers (Merton 1973a; 1973b), the work asked and inquired about the role played by context and cohort in shaping ambition, ambition thus theorized as key to realizing the goals of science at both institutional and individual levels.

In general, the foundational study found that ambition proceeds along three paths corresponding to the academic worlds. Among elites, ambition was sustained across cohorts. Among pluralists, ambition lessened across cohorts, though remained sufficiently present to enable significant research over substantial portions of careers. Among communitarians, ambition was severely modified across cohorts, this world least able to sustain what for many scientists entering this world were strong commitments and identifications with science.

As the continuum in figure 1 suggests, there were also deviations from these three normative pathways. A subset of communitarians mirrored elites in ambition, and vice versa. Both their objective records of performance and their subjective ways of accounting for their careers resembled one another. Subsets of pluralists in their ambition mirrored elites on the one hand and communitarians on the other. Here, too, resemblances among contexts could be observed in objective and subjective careers, at least among the smaller minorities of individuals who composed these various subsets.

Moving from the foundational to the longitudinal study, the research question and theoretic focus shifts. The shift allows a broader view of careers. I move from a theoretic concern with ambition to a theoretic concern with aging and adaptation. In light of the foundational work, and in ways as relevant to a study of careers in science as to careers in numerous lines of work, the question becomes fundamentally: what subsequently happens to people studied at one point in time, with their professed contextualized outlooks and temporally situated ways of accounting for themselves? More specifically, do these academics get where they wanted to go, and if not, why not? Does one reconcile obstacles confronted at the time when last interviewed or encountered in the interim span of time, and with what consequences to themselves, their careers, and the institution of science? Are they satisfied? Answers to the questions conceivably allow one to address how people's outlooks and orientations to work change and/or remain the same. One is able to more clearly see, therefore, how cohorts of people age in relation to their work.

The present means of classification aims to capture distinctions in the qualitative character of faculty work life and the ways in which academics experience and understand their careers. Such distinctions are usually obscured by other major extant means of classification, most notably the Carnegie classification of institutions (Hermanowicz 2005). For example, the 1994 version of the Carnegie classification includes all colleges and universities in the United States that are degree-granting and accredited. The 1994 classification divides institutions into six types: research, doctoral, master's, baccalaureate, associate, and specialized. It further subdivides

TABLE 5. Institutional classification systems

Institution	Academic World Classification (developed in present study)	1994 Carnegie Classification (not a part of present study)	2000 Carnegie Classification (not a part of present study)
School 1	Elite	Research I	Doctoral-Research—Extensive
School 2	Elite	Research I	Doctoral-Research—Extensive
School 3	Pluralist	Research I	Doctoral-Research—Extensive
School 4	Communitarian	Doctoral I	Doctoral-Research—Extensive
School 5	Communitarian	Doctoral I	Doctoral-Research—Extensive
School 6	Communitarian	Doctoral II	Doctoral-Research—Intensive

each of the first four types (research through baccalaureate) into two tiers (e.g., research I, research II). The system of classification is based on one criterion: an institution's highest level of degree conferred and the percentage of this degree among all other degree types awarded. The technical report states that this criterion is used in order to classify institutions "according to their missions" (Carnegie 1994, vii). Yet it is apparent that even institutions of one type, for example, research I universities, differ significantly in their missions and organizational reward structures. Moreover, authors of the Carnegie report wrote that they "oppose the use of the classification as a way of making qualitative distinctions among the separate sectors" (Carnegie 1994, vii). Yet it is precisely these qualitative details that are of concern, since they likely reveal significant differences among institutions and the organizational conditions for careers.

The Carnegie classification underwent a revision in the year 2000, but the results are even more ambiguous. In this iteration, institutions are grouped by eliminating the I and II tiers. Consequently, institutions are rendered more similar than different, directly at odds with the goals of the present research and indeed with any research that attempts to uncover significant and meaningful contrasts among the institutions that comprise the system of American higher education.

To place the problem in greater relief, a comparison of classification systems is made for the institutions composing the present sample, as depicted in table 5. In the table, the six institutions whose physics departments were sampled in this work are listed along with the academic worlds, following the logic of moral orders, *used herein*. Also listed in the table are the Carnegie classifications, *not used* in the present work, but which would correspond to the institutions in the study.

The 1994 Carnegie classification makes the institutions in the present work appear highly alike. The only difference among them captured is the sole criterion on which the system is based—the kind and percentage of highest degree conferred. Instituted under the guise of progress, the 2000 Carnegie classification masks distinctions even further. The three sets of highly different institutions under present study appear essentially the same. Thus the utility of the present classification is found in its capacity to capture those distinctions that meaningfully locate academic worlds in the practices, beliefs, and career performance patterns of the practitioners employed in them.

RESEARCH DESIGN AND SAMPLE

Nearly all of the scientists interviewed originally in 1994 worked at the same institutions when I reinterviewed them ten years later. Only four of the scientists had relocated (the reasons for which I will discuss below). This is both a finding in its own right and, more important, a methodological point that informs the study's design. Mobility among institutions is notably low throughout the sample, and the sample—which also will be discussed momentarily—includes scientists whose records are sufficiently prime as to make them comparatively easy candidates to move either on their own volition or through the recruitment initiatives of competing institutions. While a subset of these scientists may have entertained thoughts or offers to move, they have not done so, for the most part.

Methodologically, the lack of mobility means that the original study design remains essentially intact. This may be seen to carry significant advantages. From a data-collection standpoint, it made the acquisition of the longitudinal data arguably easier and more efficient than if one were confronted with the situation of finding much of the original sample dispersed to new institutions and geographic locations. That scenario would have likely lengthened the time for the necessary fieldwork or rendered it impracticable altogether because of substantially increased costs. As such, the fieldwork for the follow-up study at the original institutions occurred with an ease probably not to be found if one doubled or tripled the number of institutions.

More significant, from a theoretic standpoint, the lack of mobility enables us to examine individual development and the structure of careers under a constancy of institutional contexts. This is not to say that those contexts are unchanged in the ten-year period, for indeed some have changed in significant ways, as I will indicate. Rather, this means that one is in a position to study how durable work contexts influence and affect individuals over time. To be sure, it is also within the purview of the present concerns—as they have been defined—to investigate how change in one's institutional location might affect a career and the individual attempting to develop it. This will be less possible to accomplish given the scientists' low institutional mobility, though the four cases where this has occurred will lend some insight into the reasons why mobility occurs, when in the career it may be most apt to happen, and what consequences these changes have on individuals, their scientific work, and their professional outlooks. By the same token, one cannot study scenarios of mobility that do not exist empirically, and I thus focus on what the data suggest about careers, as experienced by those who establish the study's empirical base.

Like the present sample, the mobility rate of the population of tenured or tenure-track faculty across academic fields is relatively low. Using the National Survey of Postsecondary Faculty of 1999, in which more than 19,000 faculty members were sampled across 959 institutions, Jack Schuster and Martin Finkelstein (2006, 217) report that 32.2 percent of all regular faculty held no academic job prior to the one they currently hold. Slightly over 31 percent of all regular faculty held one previous academic job, but of this fraction, only 15.1 percent of these jobs were in full-time, tenure-line positions. The percentage of regular faculty members having held two previous academic jobs is 36.7, but of this fraction, only 10.9 percent were in full-time, tenure-line positions.[2] Thus, while there may be mobility, for most academics it does not occur among tenured or tenure-track positions. For most academics in tenure-line positions, their first appointment will be the only one they have in the course of their academic career.

Table 6 presents the study's longitudinal research design. Based on the foundational study's design depicted in table 4, the follow-up design includes the six universities that originally composed the frame from which to sample departmental physics respondents. The six universities are in turn codified into the three institutional prototypes discussed above—elite, pluralist, and communitarian. The design further includes a differentiation by professional age/career phase, resulting in three professional cohorts, as indicated by the year in which the scientists earned their PhDs: prior to 1970; between 1970 and 1980; or after 1980.

TABLE 6. Research design and number of scientists in longitudinal study

Institution	Cohort (by year of PhD)			
	Pre-1970	1970–1980	Post-1980	Total
Elite				
School 1	8	6	6	20
School 2	1	—	2	3
Pluralist				
School 1	5*	4*	6*	15
Communitarian				
School 1	3	3	3	9
School 2	1*	2	2	5
School 3	1	—	2	3
Total	19	15	21	55

Note: Asterisks indicate the cells where attrition occurred between first and second studies.

An asterisk identifies those cells in the design where subject attrition occurred. An examination shows this occurred infrequently and then only in four of the cells: one individual for each of the three cohorts in the pluralist institution, and two individuals of the eldest cohort in one of the communitarian institutions.

Whereas the response rate by participants in the foundational study was 70 percent, it was 93 percent in the longitudinal study. The attrition from the original sample is depicted in table 7. Just one of the scientists had passed away, in 1999. Four other scientists declined to be interviewed for the longitudinal study. Among these four, the reason offered for not participating in the longitudinal work was most frequently "lack of time." Thus, of the 59 available scientists who composed the original sample, 55 of them agreed to participate once again and form the basis of the longitudinal work.

The cells in which institutional mobility occurred from the time at which scientists were first studied in 1994 to when they were restudied from 2004 to 2005 are presented in table 8. It is notable that in all four cases of its limited occurrence, institutional mobility is found only in the youngest cohort of scientists. Of the four, three were from elite institutions, one from the pluralist institution.

TABLE 7. Voluntary and involuntary attrition of scientists in longitudinal study, by cohort and academic world (involuntary attrition in parentheses)

Cohort/ Career Stage	Elite	Pluralist	Communitarian	Total
PhD pre-1970		1		1
PhD 1970–1980		1		1
PhD post-1980		(1)[a]	2	3
Total		3	2	5

Note: [a] Deceased, 1999.

TABLE 8. Scientists who moved to other institutions 1994–2004, by cohort/ career phase and academic world

Career Phase	Cohort	Elite	Pluralist	Communitarian	Total
Early	PhD post-1980	3	1		4
Middle	PhD 1970–1980				
Late	PhD pre-1970				
Total		3	1		4

TABLE 9. Institutional types to which scientists moved and reasons for moving

Case	From	To	Reason
1	Elite	Pluralist	Tenure Denial
2	Elite	Pluralist	Better Job
3	Elite	Pluralist	Pre-Tenure Ambivalence
4	Pluralist	Elite	Administrative Role

What became of these scientists? Table 9 shows where they went and the reasons for doing so. In all four cases, the scientists took other academic positions. The three scientists previously employed at elite institutions moved to pluralist institutions, universities akin to the pluralist institution included in the foundational study of this work. Their reasons for doing so varied to a degree. One was denied tenure. A second had grown ambivalent about a successful tenure outcome. The third moved, in advance of tenure, for what was described by the respondent as a "better job." The one former pluralist who switched institutions moved into the elite, and did so in order to assume an administrative post. The move occurred after the individual had earned tenure as well as promotion to full professor at his former institution.

The institutions to which these scientists relocated fit within the general system of classification discussed earlier, since this system is posited to include all higher education institutions whose locations may be mapped and identified in sociocultural space. In moving, for instance, from Caltech to the University of Florida, one switches from an elite to a pluralist institution, just as, say, in moving from the University of Oregon to Harvard, one switches from a pluralist to an elite institution. While these cases of institutional mobility are few in number, they introduce variations to career paths and patterns that, when explored, offer insight into the structure and experience of academic work. I will discuss these cases in the next chapter as part of examining the career passages of the youngest cohort of scientists.

In order to eventually account for academics' unfolding career perspectives, it is useful to know about not only their present institutional locales but also their institutional origins. In particular, their doctoral-granting institutions may be recognized for socializing effects, not the least of which include the expectations people form for their future careers and norms governing achievement. The graduate institutions at which the scientists earned their doctoral degrees are listed in table 10. Insofar as their doctoral origins, scientists presently employed at elite and pluralist institutions are more similar to one another than are those employed at communitarian institutions. Fourteen of the twenty-two elites, or 63.6 percent, earned doctorates at institutions considered to have one of the top ten departments of physics in the United States.[3] This percentage for pluralists was a comparable 62.5 percent. For communitarians it was 17.6 percent.

Departments ranked in the "top ten" are, of course, but one threshold of performance. Many good, if not excellent, departments will rank outside of the top ten, this number sometimes owing itself to pure symbolism. Furthermore, some, if not many, graduates of departments outside of the top ten will possess achievement expectations akin, and at times identical to, graduates of the most reputed departments. If one applies less stringent measures, the scientists in all three worlds more closely resemble one another. For example, the percentage of scientists who received their doctorates at major research universities in the United States is high across the board. Perhaps even more telling is employment in academia in and of itself. Academic labor markets are intensely competitive, particularly in periods of

TABLE 10. Graduate institutions of scientists, by current institutional identity

Elites (N = 22)	Pluralists (N = 16)	Communitarians (N = 17)
Berkeley	Berkeley	Berkeley (2)
Birmingham	Bombay	Caltech
Caltech (2)	Cornell	Colorado
Chicago (3)	Harvard (3)	CUNY (2)
Cornell	Illinois (2)	Georgia
Harvard (2)	Indiana	Iowa State
Landau Institute[a]	Johns Hopkins	Louisiana State
London	Maryland	Minnesota
Milan	M.I.T. (3)	Missouri
M.I.T. (4)	Oxford	Nebraska
Minnesota	Pennsylvania	NYU
Northwestern		Tokyo
Pennsylvania		Wisconsin
Princeton		Wroclaw[b]
William & Mary		Virginia

Notes: [a]Russia; [b]Poland

reduced financial support for research. Many students who manage to obtain doctorates and who seek a career in academia do not manage to obtain an academic job. Thus those who successfully penetrate the academic labor market are, on balance, probably highly motivated, skilled, and productive. In general, this is probably true regardless of doctoral origin. In this light, all of the scientists, regardless of present and past institutional affiliations, share baseline attitudes and behaviors about achievement in science. These institutions, and these considerations, should be kept in mind because they will bear on discussion in later chapters about how scientists' expectations and perceptions of their careers change over time.

Four of the scientists in the sample are women. This constituted 6.6 percent of the original sample of sixty physicists (and 7.3 percent of the longitudinal sample of fifty-five physicists). As I stated in the foundational study, 5.4 percent of physics faculty members at graduate-degree-granting institutions were women (American Institute of Physics 1994, see tables 1 and 2, p. 2–3). Thus, at the time the original sample was drawn, it represented a slight *oversampling* of women physicists. The four women were arrayed across cohorts and academic worlds. One woman was in the eldest cohort, in the elite academic world. One was in the middle cohort, in the pluralist academic world. The final two were in the youngest cohort, one in the elite, the other in the communitarian, academic world. None of the women comprises a case of institutional mobility. At the time of the longitudinal study, all of the women remained employed by the institutions where I originally interviewed them. All four women were revisited as part of the longitudinal study.

THE TEN-YEAR CAREER INTERVAL

The ten-year interval used to study careers in the present work is not a random interval of time. Rather, it has both practical and theoretic importance. From a practical point of view, the ten-year mark represents a point at which the greatest number of respondents from the original sample would have been available for longitudinal study. A longer time interval would have posed significant risk of involuntary attrition. In balder terms, given the age distribution of the sample, a significant fraction of it would have been lost by natural death were follow-up work to have commenced at a later point in time. This would seriously compromise both the design and the goals of longitudinal study. On simply practical grounds, if longitudinal work was to be done, the ten-year mark appeared to be a sensible time at which to complete it.

In addition, there are theoretic reasons for choosing this time interval. Ten years of time accomplishes a major outcome: it places all of the original respondents at significantly different phases of their careers. More to the point, because the respondents were originally sampled at early, middle, and late phases of their careers, the ten-year interval advances all of them into the next set of three parallel phases. Thus the ten-year interval of time allows us to examine scientists who move from early to mid-career, scientists who move from mid- to late career, and scientists who move from late to post-career at the outermost point, but which also includes the most senior scientists who have yet to retire and/or exit their occupational careers.[4]

A five-year interval of time would not have necessarily accomplished this theoretic objective, nor would necessarily a seven-year interval, and certainly not a three-year interval. A sufficient amount of time needs to pass in order to track change and continuity, in this as in any such behavior within the rubric of temporal study. Ten years becomes a logical point at which to revisit these scientists, both because nearly all are still living and because it enables us to witness the potential of significant development in a career. Moreover, one is able to see this development (or the lack thereof) at multiple phases

TABLE 11. Number of scientists, by cohort/career phase and academic world

Cohort/ Career Phase	Elite	Pluralist	Communitarian	Total
Early to Mid	8	6	7	21
Mid to Late	6	4	5	15
Late to Post	9	5	5	19
Total	23	15	17	55

that, when combined with the foundational study, now reach from the start of a career for members of the youngest cohort to the end of a career for members of the eldest cohort.

Situated in these terms, the sample assumes a character exemplified in table 11. Original members of the early-career cohort across the three types of academic worlds have now passed into mid-career phases. Original members of the middle career cohort across the academic worlds have now passed into late-career phases. And original members of the late-career cohort across the academic worlds have now passed into the latest-career phases in several cases and into post-career phases in still several other cases.

This progression of time includes a host of experiences, including changed and evolving outlooks on work and career as well as lessons learned about a life lived in academe. Analysis of these concerns form the subsequent chapters. But a series of status changes also transpires within this time interval. Status in any line of work, including the academic profession, may be indicated in varieties of ways, but for academe a central way is professorial rank. Table 12 presents the rank advancement of the scientists between the time they were first interviewed and when they were last interviewed. The table presents the rank of each of the fifty-five members of the longitudinal sample in order to capture the numerous variations in pattern that rank advancement assumes even for this number of scientists.

Examining table 12, one notes that change in status (if not always in rank per se) is more characteristic than not for the sample, regardless of which cohort one may scrutinize. For instance, even among the eldest cohort, twelve of the nineteen scientists experienced a status change, though not a change in rank, but a change from the status of an active to a retired member of a faculty.

TABLE 12. Rank advancement of scientists, by cohort and academic world

EARLY- TO MID-CAREER COHORT (N = 21)

Case	Original Academic World	Rank @ 1st Intvw	Rank @ 10Yr FwUp
1	Elite	Asst.	Assoc.
2	Elite	Asst.	Full
3	Elite	Asst.	Chaired Full
4	Elite	Asst.	Full
5	Elite	Asst.	Full
6	Elite	Asst.	Full
7	Elite	Assoc.	Full
8	Elite	Assoc.	Full
9	Pluralist	Asst.	Assoc.
10	Pluralist	Assoc.	Full
11	Pluralist	Assoc.	Full
12	Pluralist	Asst.	Full
13	Pluralist	Assoc.	Full
14	Pluralist	Assoc.	Assoc.
15	Communitarian	Asst.	Assoc.
16	Communitarian	Asst.	Full
17	Communitarian	Assoc.	Full
18	Communitarian	Assoc.	Full
19	Communitarian	Assoc.	Full
20	Communitarian	Assoc.	Full
21	Communitarian	Assoc.	Full

MID- TO LATE-CAREER COHORT (N = 15)

22	Elite	Full	Full
23	Elite	Chaired Full	Chaired Full
24	Elite	Full	Chaired Full
25	Elite	Full	Full
26	Elite	Full	Chaired Full
27	Elite	Full	Full
28	Pluralist	Full	Chaired Full

(continues)

TABLE 12. (continued)

MID- TO LATE-CAREER COHORT (N = 15)

Case	Original Academic World	Rank @ 1st Intvw	Rank @ 10Yr FwUp
29	Pluralist	Full	Full
30	Pluralist	Full	Full
31	Pluralist	Full	Chaired Full
32	Communitarian	Assoc.	Assoc.
33	Communitarian	Full	Chaired Full
34	Communitarian	Assoc.	Retired Assoc.
35	Communitarian	Full	Full
36	Communitarian	Full	Retired Full

LATE- TO POST-CAREER COHORT (N = 19)

Case	Original Academic World	Rank @ 1st Intvw	Rank @ 10Yr FwUp
37	Elite	Full	Retired Full
38	Elite	Full	Retired Full
39	Elite	Full	Full
40	Elite	Chaired Full	Retired Chaired Full
41	Elite	Chaired Full	Retired Chaired Full
42	Elite	Chaired Full	Chaired Full
43	Elite	Chaired Full	Chaired Full
44	Elite	Full	Retired Full
45	Elite	Full	Retired Full
46	Pluralist	Full	Retired Full
47	Pluralist	Full	Full
48	Pluralist	Chaired Full	Retired Chaired Full
49	Pluralist	Full	Retired Full
50	Pluralist	Full	Full
51	Communitarian	Full	Full
52	Communitarian	Chaired Full	Retired Chaired Full
53	Communitarian	Assoc.	Retired Assoc.
54	Communitarian	Full	Retired Full
55	Communitarian	Full	Full

The greatest rate of change occurred among the youngest cohort as they advanced into middle phases of their careers. Twenty of the twenty-one members of the early cohort experienced a change in rank. Those who were assistant professors when first interviewed were now associate and, in several cases, full professors. In other cases, associate professors of the foundational study were full professors when reinterviewed ten years later.

In one illustrious case (no. 3), an assistant professor advanced to that of a chaired professor.

In purely relative terms, the middle cohort bears the least, but still a significant degree of change. Virtually half of this subsample—seven of the fifteen middle cohort members—experienced a change in professorial rank between the foundational and longitudinal studies. Five of these seven cases occurred when full professors advanced into professorial chairs. The balance of the two cases were retirements, one as a full professor, the other as an associate professor.[5]

On the basis of this table, there do not appear to be significant variations in rank advancement by institutional type, with one exception: the greater concentration of professorial chairs held by elites. Eight elites occupied a professorial chair in the longitudinal study, compared to only three pluralists and just two communitarians. One explanation for this stratification flows from how the reward system in science operates. If professorial chairs are designated for those whose work has had a particularly strong impact, and is thereby worthy of particularly notable recognition, this will most likely occur in the elite world, which both selects for and facilitates such achievement through its structure of comparatively plentiful resources and culture of outstanding role performance to advance scientific work. Another explanation for the observed stratification, though not incompatible with the explanation above, is that elite institutions simply surpass pluralist and communitarian institutions in their level of monies and endowments to create a larger number of professorial chairs for people to occupy.

As time proceeds, so of course does age. The age distribution of the sample parallels roughly these general sets of phases—early, middle, and late—that may be seen to compose a scientific career. But, even more important, advancing age is accompanied by a progression of experiences that inform what it means to be a scientist, to have an academic career, and to work in a university. In short, age offers a base from which to see how one's overall outlook on self, work, and profession change over time within a specific type of institution.

Table 13 presents the age ranges of the scientists and their average age by cohort, at both the time they were first interviewed and when they were interviewed for this study. The table also presents the range of years of experience and the average number of years as a professor by cohort, again for when the scientists were first interviewed and when they were interviewed for this study.

When first interviewed, members of the early-career cohort ranged in

TABLE 13. Age, age ranges, and years of experience of scientists, by cohort and time of interview

Cohort/ Career Stage	Age Range @ 1st Intvw	Avg. Age @ 1st Intvw	Age Range @ 10 Yr FwUp	Avg. Age @ 10 Yr FwUp
Early to Mid	32–45	37.0	42–55	47.0
Mid to Late	42–55	48.3	52–65	58.3
Late to Post	52–67	61.4	62–77	71.1
Cohort/ Career Stage	Range Yrs. Experience @ 1st Intvw	Avg. No. Yrs. as Prof. @ 1st Intvw	Range Yrs. Experience @ 10 Yr FwUp	Avg. No. Yrs. as Prof. @ 10 Yr FwUp
Early to Mid	2–9	5.0	12–19	15.0
Mid to Late	9–24	16.3	19–34	26.3
Late to Post	21–38	30.2	31–44	37.0

age from 32 to 45 and had between 2 and 9 years of experience as professors. In comparison, members of the middle career cohort ranged in age from 42 to 55 and had between 9 and 24 years of experience as professors. Members of the late-career cohort ranged in age from 52 to 67 and had accumulated between 21 and 38 years of experience as professors. Adding 10 years to each of these sums, age advances with experience, and future discussion will reveal how the scientists come to view themselves and their work in light of this progression.

For a subset of the sample, this progression takes scientists out of their careers entirely, or at least out of formal employment from the universities where they have spent substantial fractions of their professional lives. As table 14 indicates, one-quarter of the sample (25.5 percent) had retired as of the time the longitudinal fieldwork was conducted. Of the fourteen scientists included in this percentage, twelve were in late phases of their careers, as would be expected; the two others who retired came from the mid- to late-career cohort.

Together, these individuals present the opportunity to investigate two fundamental concerns about academic careers. First, how do scientists make the passage into retirement? How is the passage perceived, and with what consequences? Second, given that their careers are in some sense "completed," how do scientists view their careers, their work, and their lives, which have—in varying ways—been devoted to work and career? In short, how does it all look at the end? And do outlooks on the career at

TABLE 14. Retired scientists as proportion of sample, by cohort/career stage and academic world

Cohort/ Career Stage	Elite Retired/ Sample	Pluralist Retired/ Sample	Communitarian Retired/Sample	Total
Early to Mid	0/8 (0%)	0/6 (0%)	0/7 (0%)	0/21 (0%)
Mid to Late	0/6 (0%)	0/4 (0%)	2/5 (40.0%)	2/15 (13.3%)
Late to Post	6/9 (66.6%)	3/5 (60.0%)	3/5 (60.0%)	12/19 (63.2%)
Total	6/23 (20.6%)	3/15 (20.0%)	5/17 (29.4%)	14/55 (25.5%)

its end differ significantly from those formed by scientists not yet there, in earlier career phases? To answer these questions, and the larger research questions from which they derive, data is needed from those whose careers and career perspectives can inform the issues at hand.

THE FIELDWORK

There were only two points of contact I had with the respondents since first interviewing them in 1994. The first was through a letter I sent them immediately upon completing the interviews, acknowledging their help and participation. The second was a letter I sent in August 1998, informing them of the foundational study's results and its publication. This fulfilled the promise I had made at the time of the first interviews to inform them about the research. But there was no further correspondence. In the interim, I had read about the accomplishments of several of the scientists in the sample in various venues, such as the widely circulating journal *Science* and national newspapers, such as the *New York Times*. There were no iterations of subsequent data collection until the ten-year longitudinal project. The foundational study was conceived and completed in the absence of entertaining a possibility of longitudinal work. The subsequent discovery that such an undertaking would be the first longitudinal study of the academic profession that followed the same subjects over time suggested that this might be a profitable new way to examine careers.

Since the ten-year longitudinal study was designed to examine the experience of academic careers and the outlooks that scientists form as their careers unfold, the method for the study called once again for a capacity to differentiate among fine-grained meanings and to obtain a level of detail that would inform what it means to be a scientist over the course of a long career. The interview was thus retained as the primary means of data collection.

I began contacting respondents of the original sample in Spring 2004. I did so by sending them the letter that appears in appendix B. The letter attempted to place the longitudinal study in context by reminding them of their previous participation in the foundational work, by explaining what the longitudinal study sought to accomplish, and by informing them about what their continued participation would involve. Subsequently, I telephoned each of the potential respondents. At this point, I described the project more fully, invited any questions they might have, and ultimately asked if I could meet with them again for an interview. If they agreed, interview arrangements were made, and the scientists were also asked to provide a copy of their curriculum vitae at the time of the interview.

Following the interviews, I sent the respondents another letter to thank them for their participation, their time, and their insight. By way of example, this letter appears as appendix C. The letter also presented the final opportunity to give the scientists my address and telephone number, should any of them wish to contact me at a subsequent point in time. I also included in this letter the contact information for the human subjects officer who oversaw the human subjects review of this project at the university where I work and whom research participants were free to contact if they had questions they believed should be posed to this individual.

The 93 percent response rate is indicative of a high level of cooperation, but it does not reveal the high level of warmth and enthusiasm I encountered in reestablishing contact with the scientists. When I made telephone contact with the scientists, they were highly agreeable and receptive to another meeting. "Of course I would be willing to meet with you again," "Sure, I'd be happy to see you," "Yes, when were you planning on being here?"—these were the customary responses that met me, even as I tried to arrest concerns of attrition, which I knew would irreparably compromise the project.

Roughly a week prior to each of the interviews, I sent e-mail to the scientists to remind them of the date and time of our meeting. Unless the arrangements had to be changed (which occurred on only one occasion), there was no need for a reply. Nevertheless, several scientists did reply, and the contents bespeak characteristics of the population with which I am dealing. One scientist wrote back (in the informal, all-lowercase vernacular of many e-mail writers): "yes, i have you down on my calendar, and look forward to seeing you here. my address is ——, though the name tag is not on the door, and the phone is —— ." Another scientist wrote back: "Funny, I've been e-mailing people in physics and chemistry at your University all morning. In any event, I have indeed remembered the interview time and have printed out a c.v. already." All the scientists were cordial. Several went out of their way in helping make arrangements. They provided elaborate directions to campus and directed me to specific parking lots and the buildings and offices where I could find them. Several scientists e-mailed me maps of their university and made provisions through departmental secretaries for parking passes, which they had waiting for me upon my arrival. Many asked where I would be staying when meeting with them. Several others invited me to lunch, which, because of my interview schedule and because of a desire to maintain a deference to the interview and its procedures, I regretfully declined.

I was struck by the degree of this generosity, but I have found scientists a relatively easy population to study, insofar as fieldwork is concerned. For many scientists, research is a staple of their professional lives. This may dispose them to unusually high levels of helpfulness in others' research in which their participation is crucial. What is more, in their many roles, scientists are called to speak about their work. They thus develop trained ways of describing what they do and how and why they do it, if not always to general audiences, then to scientifically trained ones.

Academic scientists are also situated in a clearly defined career, with delimited ranks and general understandings of rank durations, all geared to the idea of progress and advancement, both for the sake of science and for a career. This likely brings about a consciousness and reflectiveness about one's work, career, and progress perhaps not found to such a degree in many other lines of work, especially non-professions. In general, scientists appear to enjoy the opportunity to talk about themselves and their work and therefore seem especially compatible (as other academics arguably would) with this type of study. Thus, while the response rate in this work is high regardless of the way one looks at it, it may be explained by the characteristics of the group under study.

The interview protocol was divided into five parts and appears as appendix D. The parts of the protocol were guided by the three theoretic perspectives of the work discussed in the introduction: an occupational perspective, a life-course perspective, and a sociology of science perspective. In the study, the perspectives converge to form a focus on aging in relation to work. Interview questions were thus designed to examine people's perspectives on their careers.

Part one of the protocol sought to inquire explicitly about changes and continuities in the scientists' work-lives. The major object of inquiry was scientists' career perspectives and how they believed these had changed over time, particularly over the past ten years. I asked the scientists about any significant changes outside of their lives at work, about whether they believed their careers had progressed as anticipated, and whether they perceived any changes in their intensity toward work. To inquire further about continuities and changes, I asked them about how they perceived younger and older careers to differ, as well as how careers might differ in different types of institutions. I asked if they had seen their definitions of success change, given the centrality of achievement and success to most scientists' careers. Finally, I asked what change they would *want* to see made if, perchance, they were a university president and held the power to bring about change.

Part two of the protocol inquired about scientists' satisfactions. The purpose of these questions was to ask what were the chief satisfactions for these scientists. I also asked when in their careers, and what about this point, scientists were the most satisfied. Finally, I asked the scientists, if they were to start all over again, whether they would still seek an academic career.

Part three of the protocol asked the scientists about their dissatisfactions, and as such was designed to complement—and thus more fully elucidate—data provided in response to the preceding questions about satisfaction. I specifically inquired about what in their careers the scientists wished they could do differently. I asked about possible frustrations and about their perception of the reward system in science. I asked them if they had found an academic career to be unrewarding in any way, and whether they had ever entertained the possibility, in any serious way, of leaving their present institutions. Finally, I asked them to indicate what they have come to perceive to be the greatest weakness of their graduate training, the expectation being that answers to this question would both identify dissatisfactions and pinpoint methods they use to deal with these dissatisfactions.

Part four of the protocol asked scientists about their aspirations. I inquired about how the scientists perceived their aspirations to have changed, particularly within the last ten years, in order to understand how they evaluate the progress they had made toward goals, especially those they articulated as important to them when last interviewed. Finally, for those scientists who had retired, I asked what they missed most about their jobs. For all others, I asked if they would retire now if given the opportunity to do so while leading the same quality of life.

Part five of the protocol contained questions specifically about retirement. It contained two main questions that were asked only of retired scientists. The questions dealt with the best and worst aspects of retirement, as perceived, and about adjustments scientists saw themselves having to make in their passage into retirement. The questions also provided an opportunity for scientists to discuss what retirement variously meant to them.

I asked a final question of all of the scientists, namely what they took to be an accomplishment of which they were most proud. I asked this question in order to conclude each interview on an unambiguously positive note, which adheres to standard conventions of interview protocol design. Such design is especially important in interviews that cover sensitive topics or that have asked respondents to speak candidly about issues at prior points in the protocol that may cause anxiety or discomfort.

Of the fifty-five interviews, forty-eight were conducted in person in the scientists' departmental offices. The balance of the seven interviews were conducted by telephone, an arrangement that arose out of necessity: in most of these cases, the scientists were not in town when I visited their departments and alternative interview arrangements had to be made. In other cases, scientists had moved and were distant from the institution where I had interviewed them previously. It was not feasible to travel to faraway sites for single interviews. In the four cases in which scientists moved because of new institutional locations, I interviewed three in person. In two cases, the scientists were within reasonable distance from either my home institution or other interview sites, and in one case a scientist had relocated to a department that had been included in the foundational study and was thus interviewed when I revisited that institution.

At the interviews, I obtained copies of each scientist's curriculum vitae.[6] In academic and scientific careers, the curriculum vitae is a standard document that lists the positions and activities of the academics who ritually compose them, both to serve as an official record of their roles and for the more mundane purpose of review of role performance conducted by the departments, colleges, and universities where individual academics hold rank. I obtained the vitae in order to track scientists' positions and activities, including their publication productivity. Each vita thus becomes a form of data that can be analyzed, which was the case in the foundational study when each scientist's CV was first collected. In the present longitudinal work, continuities and changes in positions and activities may be observed over time in these documents in order to glean patterns of academic careers across cohorts and contexts of science.

At the conclusion of each interview, I asked the scientists to complete a brief questionnaire (administered by myself in the case of the telephone interviews). The questionnaire obtained data on three specific demographic and personal items: marital status; the number and age of any children; and salary. These data were obtained to further contextualize the accounts scientists provided, thus allowing one to see, for example, whether significant variations in scientists' experiences and work understandings are associated with these general demographic characteristics. Like the CVs, data from the post-interview questionnaire also represent a basis on which to draw comparisons and contrasts from findings of the foundational study, where a similar questionnaire was administered. The post-interview questionnaire is presented in appendix E.

Finally, I collected data on each of the six departments of the foundational study. These data are also used to form the core of the longitudinal work. I mailed to the head/chair of each of the physics departments a questionnaire that sought to collect data on the major infrastructural conditions of work in those departments. The questionnaire paralleled one I administered for the foundational study in which I similarly asked department heads/chairs to supply information about their departments. The questionnaire obtained data on teaching and resources available from the departments and their respective universities. Like all the other sources of data used in this study, the questionnaire of the longitudinal work will be compared with its foundational counterpart in order to examine departmental continuities and changes.

In addition, the departmental questionnaire used in the present longitudinal study introduced an explicit section on "Department Change." Here information was sought from heads/chairs about conditions of work that have evolved in the ten years since the foundational work. The purpose of the questionnaire was to establish important objective conditions of work under which careers are subjectively experienced and understood. If departments differ substantially in the objective conditions under which careers transpire, this likely entails substantial consequences in how individuals subjectively construe their work-lives and in how their careers unfold. Like the post-interview questionnaire, data from the questionnaire given to departments will be used to further contextualize the career accounts provided by the scientists. The departmental questionnaire is presented in appendix F.

A list of the sources of data used in the present study is provided in table 15. The table summarizes the three data sources used in both the foundational and in the longitudinal studies: interviews, post-interview question-

TABLE 15. Summary sources of data utilized in ten-year longitudinal study

Data Source	Foundational Study	Longitudinal Study
1. Interviews	X	X
2. Post-Interview Questionnaire	X	X
3. Departmental Questionnaire	X	X

naires, and departmental questionnaires. It also highlights the methodological point made throughout the preceding discussion that each of these data sources will be incorporated to form the present work.

ANALYSIS OF DATA

All of the interviews were conducted by the author. The interviews averaged sixty minutes in length. All were tape-recorded and transcribed. The interviews constitute more than 1,700 pages of double-spaced transcript. The transcripts in turn were coded by the author to facilitate analysis of the data.

The specific codes employed pertained to the subject headings of the interview protocols, in both the foundational and longitudinal studies. For example, a question under the longitudinal study protocol section "Changes and Continuities" asks: "Do you think you are working harder, less hard, or about as hard as you were ten years ago?" Responses were coded using the same response categories offered in the question. Under the section "Satisfactions," a question asks: "In learning what you have about academic careers, would you go into an academic career if you were starting all over again?" In this example, responses were coded affirmatively or negatively, and the probe question was coded for the explanation provided for the response, using codes such as "funding," "difficulty," "lack of reward," and "freedom."

In coding and analyzing the longitudinal data, I paid particular attention to how responses coalesced around themes of consistency and change. Following Saldana (2003, 64), I employed a variety of conceptual and thematic questions to help situate data analysis, including:

1. What increases or emerges through time?
2. What is cumulative through time?

3. What kinds of surges occur through time?
4. What decreases or ceases through time?
5. What remains constant or consistent through time?
6. What is idiosyncratic through time?
7. What is missing through time?
8. Which changes interrelate through time?
9. What are participant or conceptual rhythms through time?
10. What is the characterization of across time experience, and how do characterizations differ by sub-groups of the sample?

In all instances, my intent was to formulate understandings of respondents' experiences and to derive substantive comparisons and contrasts with respect to the key dimensions of the research design: the institutional contexts in which scientists worked and the temporally situated cohorts of which they were members. This allows me to address the guiding question of how scientists age in their work environments.

I utilized an approach to data analysis most often referred to as "constant comparison" (Charmaz 1990; 2001; Glaser and Strauss 1967). This approach is marked by several characteristics. A researcher simultaneously collects and analyzes data. In the course of doing so, the researcher pursues emergent themes and begins to discover basic social processes in the data. These themes and processes are elaborated, modified, or qualified through further data collection and analysis. The researcher inductively constructs and refines abstract conceptual categories that explain and synthesize these themes and processes. The researcher integrates categories into a meaningful theoretic framework that specifies conditions and consequences of the studied processes (Charmaz 2007; Charmaz and Mitchell 2001).

Through constant-comparative analysis, typical, predominant patterns are gleaned from the data. These patterns are what statistically would be referred to as central tendencies. I used such a comparative method for all nine cells constituted by the three types of institutions crossed with the three professional age cohorts. My objective was to determine, insofar as possible, the modal experience of work by individuals located in each cell of the research design.

To supplement this analysis, the task then turns to "deviating cases," or what others sometimes call "negative cases," which may be defined as those cases departing from the typical found in any given sub-grouping (Charmaz 2001). As Becker and his colleagues explained in their illustrious study of the socialization of medical students:

If we were to carry on our analysis by successive refinements of our theoretical models necessitated by the discovery of negative cases, we wanted to work in a way that would maximize our chances of discovering those new and unexpected phenomena whose assimilation into such models would enrich them and make them more faithful to the reality we had observed. (1961, 24)

No grouping is perfectly uniform; one may always observe some degree of variation. Where such deviating cases exist, one attempts to answer the questions of why and how they have come to depart from the norm. This procedure allows the researcher to strengthen assertions and to qualify suggestive conclusions about patterns indicative of groups and sub-groups in a sample.

This type of comparative analysis, sometimes called small-N comparison (Abbott 2004), ethnographic revisits (Burawoy 2003), or biographical approaches to the study of lives and careers (Atkinson 1998; Bertaux and Kohli 1984; Clausen 1998; Denzin 1989), proceeds within established conventions of social science research (Cairns, Bergman, and Kagan 1998; Giele and Elder 1998). This type of analysis, and variations of it, more generally owe their origins to what is known as "grounded theory," an inductive process of concept- and theory-building that transpires as data points are gathered and analyzed (Glaser and Strauss 1967; Strauss and Corbin 1994).

A version of this comparative method dates to Durkheim's classic study of suicide ([1897] 1951). In that work, Durkheim sought to find patterns that coalesce into types of suicide and in turn to observe the conditions under which certain cases deviate from their respective norms. Not each and every case of suicide may be locatable as following an anomic, fatalistic, egoistic, or altruistic type, though these types account for much of the variation in suicide that Durkheim observed. A minority of cases may deviate from these respective central tendencies and assume the characteristics of more than one type, what Durkheim called "mixed forms."

Contemporary social-scientific work similarly employs versions of this technique. Daniel Levinson (1978) used a sample of 40 men between the ages of 35 and 45 to investigate life stages; he did so again in a separate study (Levinson 1996) using a sample of 45 women to research stages of women's lives. George Vaillant (1977) used a sample of 95 men, focusing in particular on 49 cases, to investigate styles of adaptation to life events over time. Robert Weiss followed a sample of 68 widows and widowers to understand bereavement, reactions to trauma, coping and recovery patterns

(Parkes and Weiss 1983). Robert White (1966) concentrated on a sample of 3 individuals to trace the growth of personality from adolescence to adulthood. John Laub and Robert Sampson interviewed 52 men, first researched by Sheldon and Eleanor Glueck beginning in 1939, about persistence and desistence in criminal careers (Laub and Sampson 2003). As one can see, regardless of the specific research topic, the sample sizes in intensive interview studies is small compared to survey research, but sufficiently large to satisfy the declared intentions of the studies.[7]

Survey research, drawing on large samples, seeks to produce general statements about empirical regularities found within large populations (e.g., Blau and Duncan 1967). By contrast, interview research seeks to reveal people's interpretations of forces that change or reproduce social processes (Ragin 1987; Weiss 1994). Interview research typically relies upon smaller samples, which more readily facilitate in-depth inquiry, permitting researchers and readers to understand the finer-grained meanings that people assign to their lives or some aspect of their lives. While the results of interview research direct researchers toward uncovering social processes by examining the details of individually lived experience, the social processes uncovered and the theoretic categories used to explain them may pertain to more general populations. Andrew Abbott makes the following point:

> Small-N comparison attempts to combine the advantages of single-case analysis with those of multicase analysis, at the same time trying to avoid the disadvantages of each. On the one hand, it retains much information about each case. On the other, it compares the different cases to test arguments in ways that are impossible with a single case. By making these detailed comparisons, it tries to avoid the standard criticism of single-case analysis—that one can't generalize from a single case—as well as the standard criticism of multicase analysis—that it oversimplifies and changes the meaning of variables by removing them from their context. (Abbott 2004, 22)

Thus the present work does not pretend to offer a definitive statement; it offers suggestive insights into the patterned ways in which people experience work and career.

A preponderance of new questions were used in the protocol of the longitudinal study. Only the final question, about scientists' perceived greatest achievement, is common to the protocols of both studies, and this is because many of the questions asked of scientists in the foundational study were time-bound (refer to appendix A). That is, the majority of questions and responses to them were such that they provided insight when being posed at just one point in time. Overall, the questions of the foundational study have low utility value in being asked again in longitudinal work. Examples of the questions asked when the scientists were first interviewed include: "How did you come to arrive at this university?" and "What aspirations did you have as a graduate student?" and "How have your aspirations unfolded since being a graduate student?" Responses to these and other questions made for an entire study. In general, however, they are questions that, if asked again at successive points in time, would yield little in the way of empirical and theoretic advances over what had been learned through the foundational work. It is likely that most scientists' responses to these and related types of questions would be essentially unchanged over time, and thus not worth asking more than once.

It is not the interview questions per se, but the research findings of the foundational study that matter most for the present work. The foundational study established how scientists had perceived their careers up until that point in time. That study established scientists' perceptions about where they had been professionally, where they were, and where they wanted to go.

I examine continuity and change through two means. The first consists of the longitudinal interviews in and of themselves. As framed, these interviews are a means to explore how individuals perceive having changed in their careers. The second means of analysis consists of comparisons in how people perceive their careers between the foundational and longitudinal studies. The question of how scientists perceive their careers is the common ground on which the present study builds in order to examine aging in work. One is in a position to examine how outlooks on self, work, and career evolve, particularly in light of the ways in which individuals account for the shifting perspectives they develop as they age in relation to work. Moreover, owing to the design of the research, one is able to investigate individual continuities and changes within and between the three professional cohorts, as well as across the three prototypical contexts of scientific work.

Finally, a note about the validity of individuals' career accounts: *validity* may be defined as the correctness of a description, conclusion, explanation, interpretation, or some sort of account; it is the extent to which a method of data collection has the quality of being sound or true, as far as can be judged.

As I observed in the foundational study, I am engaged in meta-interpretation, interpreting how others interpret their passage through time in an academic career. The objective is not to assess how accurately individuals recall their pasts—that is impossible—but to see how individuals account for themselves and "how they organize their complex pasts to present a coherent self-identity" (Hermanowicz 1998, 42). The reader should view the accounts presented as constructed representations of and by individuals. Objective features of accounts may be checked through records, such as a respondent's curriculum vitae. Subjective features of accounts are stylized self-presentations that are, or strive to be, internally valid. What people say, particularly in the case of an interview, which has been socially defined as an occasion to speak candidly about oneself in the midst of probe questions that compel candor, has to make sense to both interviewee and interviewer. Accounts that are internally consistent are valid: it is how a respondent goes about understanding and presenting him or herself to others in a fashion that is persuasive and believable to speaker and audience alike. "The physicist Leo Szilard once announced to his friend Hans Bethe that he was thinking of keeping a diary: 'I don't intend to publish it; I am merely going to record the facts for the information of God.' 'Don't you think God knows the facts?' Bethe asked. 'Yes,' said Szilard. 'He knows the facts, but He does not know *this version of the facts*' (Dyson 1979, xi). Or, as Tamotsu Shibutani has observed in sociological terms:

> A perspective is an ordered view of one's world—what is taken for granted about the attributes of various objects, events, and human nature. It is an order of things remembered and expected as well as things actually perceived, an organized conception of what is plausible and what is possible; it constitutes the matrix through which [people] perceive [their] environment. The fact that [people] have such ordered perspectives enables them to conceive of their everchanging world as relatively stable, orderly, and predictable. (Shibutani 1955, 564)

Put in more bald-faced terms, it does not matter whether accounts are true in some natural-scientific sense. Respondents create them for specific purposes, in this case to construct and present an identity that is meaningful and, following Linde (1993), coherent for themselves and others (see also Bjorklund 1998; Cohler 1982; Cohler and Hostetler 2003; Martin 1982; Rubin 1986; Wells and Stryker 1988).

Thus, in the present work, I focus on *within-individual* continuity and change. The study emphasizes the experience and meaning people assign to their work. To that end, I will search out the ways in which people account for how they view their unfolding careers, with perhaps changed outlooks, orientations, motivations, commitments, and satisfactions over time.

ACADEMIC WORLDS THEN AND NOW

In following the scientists, one also follows the departmental and university environments where they work and within which their careers are institutionally situated. One does so not out of idle interest but because, recalling the discussion of Hughes in the last chapter, individuals and institutions are reciprocally bound to and interactively generate the other. To speak of individual careers, therefore, is to identify more general patterns prompted and produced by the institutional environments in which careers are collectively shaped.

A host of objective work conditions impinge upon the ways in which a career may be subjectively experienced. All such conditions constitute forms of resources. These in turn can be converted—in greater or lesser degrees—into outcomes that impinge directly on the objective fates and subjective appraisals of those who do, or do not, have them at their disposal. In the present case, work conditions may be fiscal, such as the amount of money available to support research. They may be physical, such as the equipment and facilities necessary for advanced work. They may be human, such as the number and quality of colleagues on whom individuals can draw for specialized expertise, or the number and quality of students who are taught and join in the research and scholarly endeavors of a faculty. Whether fiscal, physical, or human, resources are differentiated by both quantity and quality and form a crucial basis by which careers are bound.

At the time of the foundational study, I reported the "conditions of practice" that prevailed in each of the six departments that composed the study. These conditions differentiated the departments from one another and provided a further empirical basis on which to draw distinctions among the groupings of broader academic worlds. For comparative purposes, that data are reproduced in table 16.

As part of the longitudinal study, data on these conditions of practice were updated by way of the departmental questionnaire administered to department heads/chairs. The results are reported in table 17. Change in departments is evident in comparing the two tables. But while all of the departments registered various changes, one phenomenon remained the same, namely, the ways in which the three academic worlds are stratified on numerous dimensions. If one were to select at random any one of the

TABLE 16. Conditions of practice at foundational study

	Elite		Pluralist		Communitarian	
	School 1	School 2	School 1	School 1	School 2	School 3
FACULTY CONDITIONS [a]						
Department size [b,c]	~75	~75	~50	~25	~15	~15
% with PhDs from top-10 physics departments [c,d]	78%	74%	56%	16%	0%	1%
% with ext. funding [e,f]	~50%	~65%	~50%	~30%	—[g]	—[g]
% with grad. research assistant [h,i]	76–100%	—[j]	51–75%	26–50%	—[k]	1–25%
Reasons for no grad. research assistant [i,l]	no active research	—[j]	various reasons	no funding	—[k]	no funding
Number of postdocs [i]	16–20	—[j]	>25	6–10	—[k]	1–5
Number of technical support staff [i]	21–25	—[j]	16–20	—[m]	—[k]	1–5
Yearly teaching load (no. of courses) [i,n]	1–2	—[j]	1–2	2–3	—[k]	3–4
Teaching leaves? [i]	yes	—[j]	yes	no	—[k]	yes
Leave provisions [i]	negotiate with dept. head	—[j]	sabbaticals[o]	none	—[k]	sabbaticals[p]
STUDENT CONDITIONS						
No. of undergraduates [q,r]	~5000	~25000	~35000	~15000	~8000	~15000
Avg. SAT scores [s,t]	680	570	515	480	485	485
No. of physics graduate students [c]	~300	~300	~75	~35	~15[u]	~5

	575(min.)	740(med.)	600(min.)	600(avg.)	580(avg.)	550(min.)
Dept. GRE scores [c,t]						
Avg. time to physics PhD(years) [f]	~7.0	~7.0	~7.5	8.5	—[g]	—[g]
% getting jobs at PhD schools [v,w]	~60%	~60%	~50%	~45%	—[x]	—[x]
% getting jobs, general [w,y]	~60%	~75%	~45%	~60%	—[x]	—[x]
OTHER INFRASTRUCTURAL CONDITIONS						
Annual fed. support to dept. [i]	~$70M	—[j]	~$7M	~$800K	—[k]	~$1M
Annual nonfed. support to dept. [i]	—[m]	—[j]	~$1M	~$450K	—[k]	0
Annual dept. operating budget [z,i]	~$10M	—[j]	~$7M	~$80K	—[k]	~$60K
Assistant professor start-up funds? [i]	yes	—[j]	yes	negotiable	—[k]	yes
Start-up amount [i,aa]	$40K–$400K	—[j]	$50K–$450K	≤ $35K	—[k]	$60K
Restricted uses? [i]	no	—[j]	no	capital equip. only	—[k]	capital equip. only
Other dept. or university funds? [i]	yes	—[j]	yes	yes	—[k]	yes
Type/amount	variable	—[j]	$10K grants; $4K aid(theory); $10K aid(exper)	$5K grants; $1K–$100K (equip. grants)	—[k]	$2K grants
Travel/conference reimbursement? [i]	no [bb]	—[j]	yes	yes	—[k]	yes
Nature of provisions [i]	—[bb]	—[j]	~1 conf./year	$250 dept.	—[k]	~$500
Annual no. of outside speakers [i,cc]	>30	—[j]	>30	6-10	—[k]	26–30

(continues)

TABLE 16. (continued)

	Elite		Pluralist		Communitarian	
	School 1	School 2	School 1	School 1	School 2	School 3
PERSONNEL POLICY (SELECT ISSUES)						
Outside letters for tenure review?[i]	yes	—[j]	yes	yes	—[k]	yes
Customary 10 years ago?[i]	yes	—[j]	yes	yes	—[k]	yes
Customary 25 years ago?[i]	yes	—[j]	yes	no	—[k]	yes
Teaching as promotion factor[i,dd]	a lot	—[j]	somewhat	a lot	—[k]	a little

Notes:

[a] Information applies to the *departments* of physics, except where noted. For certain matters, reporting of data is approximated (i.e., "∼") to ensure the anonymity of institutions and individuals.

[b] Number of full-time tenure or tenure-track department faculty member.

[c] Source: American Institute of Physics (1993).

[d] Percentage of department faculty who received their PhDs from departments of physics ranked in the top ten by Goldberger, Maher and Flattau (1995). Due to ties in ranking, the list includes eleven schools: Harvard, Princeton, M.I.T., Berkeley, Caltech, Cornell, Chicago, Illinois, Stanford, University of California–Santa Barbara, and Texas. To arrive at the percentage, I included in the denominator only department faculty members who had U.S. doctoral degrees.

[e] Percentage of department faculty with outside research support.

[f] Source: Goldberger, Maher, and Flattau (1995).

[g] Not evaluated by Goldberger, Maher, and Flattau (1995) because departments did not meet criteria for assessment.

[h] Percentage of department faculty with at least one graduate research assistant.

[i] Source: Departmental Questionnaire administered at foundational study; see Hermanowicz (1998, 219).

[j] Department head did not return questionnaire. Information on Elite School 2 most closely approximates equivalent information for Elite School 1.

[k] Department head did not return questionnaire. Information on Communitarian School 2 most closely approximates equivalent information for the other two communitarian schools.

[l] In response to the following question: "Taking all the faculty members who do not have graduate research assistants, what would you say is the single most important reason?" (Departmental Questionnaire administered at foundational study; see Hermanowicz (1998, 219)).
[m] Missing data.
[n] Does not discriminate between courses and course sections.
[o] Also one-quarter reduced teaching load for special research assignments (according to department head, this is used by about 10 percent of the faculty).
[p] Also unpaid leaves of absence (which must be approved by the department and the university); in addition, reduced teaching for administering large grants, which also must be approved.
[q] Number of undergraduates in the university. (The number of undergraduate majors in physics was not available for all departments. More important, however, such figures obscure the extent of undergraduate teaching in "service" and other general courses.
[r] Source: Barron's Educational Series (1996).
[s] Source: American Council on Education (1992).
[t] The maximum score for the Scholastic Aptitude Test (SAT) and for the Graduate Record Exam (GRE) is 800 for each of their components.
[u] Includes only master's degree students; department no longer has a doctoral program.
[v] Percentage of doctorates, 1975–1979, who indicated they had made firm commitments for employment in PhD-granting institutions.
[w] Source: Jones, Lindzey, and Coggeshall (1982).
[x] Not evaluated by Jones, Lindzey, and Coggeshall (1982) because departments did not meet criteria for assessment.
[y] Percentage of doctorates, 1975–1979, who indicated they had made firm commitments for postgraduation employment (including postdoctoral positions and other appointments in academic and nonacademic sectors).
[z] Excludes faculty salaries; refers exclusively to non-grant support.
[aa] Amount depends on research program of specific faculty member.
[bb] Presumably, faculty members rely on their research grants or unrestricted funds disbursed by the department or personally cover their travel costs.
[cc] Estimated annual number of people outside the department who gave formal talks.
[dd] In response to the following question: "As best as possible, indicate the extent to which teaching performance factors into promotion decisions in your department." Answer choices: a lot; somewhat; a little; practically not at all; not at all.

Source: Hermanowicz, Joseph C. 1998. *The Stars Are Not Enough: Scientists—Their Passions and Professions*, 35–37, table 2. Data are based on Departmental Questionnaire administered at foundational study, p. 219.

TABLE 17. Conditions of practice at longitudinal study

	Elite		Pluralist		Communitarian	
	School 1	School 2	School 1	School 1	School 2	School 3
FACULTY CONDITIONS[a]						
Department Size[b,c]	~75	~65	~55	~20	~15	~20
% with PhDs from top-10 physics departments[c,d]	89%	73%	52%	21%	0%	0.5%
% with ext. funding[e,f]	96%	—[g]	90%	52%	—[h]	—[h]
% with grad. research assistant[i,j]	76–100%	76–100%	—[g]	51–75%	26–50%	26–50%
Reasons for no grad. research assistant[j,k]	no funding	no funding	no funding	no active research	weak talent pool	no active research
Number of postdocs[j]	16–20	21–25	>25	1–5	1–5	1–5
Number of technical support staff[j]	—[l]	6–10	21–25	16–20	1–5	1–5
Yearly teaching load (no. of courses)[j,m]	1–2	1–2	1–2	2–3	1–2	2–3
Teaching leaves?[j]	yes	yes	yes	yes	yes	yes
Leave provisions[j,n]	sabbaticals	sabbaticals	sabbaticals; awarded research leaves[o]	sabbaticals	sabbaticals	sabbaticals; course banking[p]
STUDENT CONDITIONS						
No. of undergraduates[q,r]	~5,000	~27,000	~35,000	~20,000	~12,000	~20,000
Avg. SAT scores[s,t]	735	625	570	530	530	520
No. of physics graduate students[c]	~250	~250	~140	~45	~20[u]	~25

	575 (min.)	765 (avg.)	725 (avg.)	710 (avg.)	615 (avg.)	550 (min.)
Dept. GRE scores [c,t]						
Avg. time to physics PhD (years) [c]	~5.8	~5.5	—[g]	6.5	—[h]	5.0
% getting jobs at PhD schools [f]	33%	—[g]	50%	26%	—[h]	—[h]
% getting jobs, general [f]	67%	—[g]	50%	52%	—[h]	—[h]
OTHER INFRASTRUCTURAL CONDITIONS						
Annual fed. support to dept. [j]	~$60M	~$12M	~$10M	~$2.5M	~$1M	~$1.5M
Annual nonfed. support to dept. [j]	~$600K	~$8M	~$2M	~$2M	~$300K	~$100K
Annual dept. operating budget [v,j]	~$10M	~$20M	~$15M	~$3M	~$900K	~$50K
Assistant professor start-up funds? [j]	yes	yes	yes	yes	yes	yes
Start-up amount [j,w]	—[g]	$300K theorists; $500K–$1M experimentalists	$250K–$1.2M	$100K–$300K	$75K	$50K–$100K
Restricted uses? [j]	—[g]	research only	no	capital equip. primarily	capital equip. only	capital equip. only
Other dept. or university funds? [j]						
Type/amount	yes variable	yes grad RA $14K equip. $30K (max.)	yes $3.5K aid (theory); $9K aid (exper)	yes small grants (dept. & coll.); >$10K grants (univ.)	no	yes $1400 (dept.) <$7500 (coll.)
Travel/conference reimbursement? [j]						
Nature of provisions [j]	no[x]	yes special cases only	yes from research funds above	no[x]	no[x]	no[x]
Annual no. of outside speakers [j,y]	>30	>30	>30	21–25	6–10	11–15

(continues)

TABLE 17. (continued)

	Elite			Pluralist			Communitarian	
	School 1	School 2		School 1	School 1		School 2	School 3

PERSONNEL POLICY

| Teaching as promotion factor[j,z] | somewhat | somewhat | | somewhat | a little | | somewhat | a little |

Notes:

[a] Information applies to the *departments* of physics, except where noted. For certain matters, reporting of data is approximated (i.e., "∼") to ensure the anonymity of institutions and individuals.

[b] Number of full-time tenure or tenure-track department faculty members.

[c] Source: American Institute of Physics (2004).

[d] Percentage of department faculty who received their PhDs from departments of physics ranked in the top ten by Goldberger, Maher and Flattau (1995). Due to ties in ranking, the list includes eleven schools: Harvard, Princeton, M.I.T., Berkeley, Caltech, Cornell, Chicago, Illinois, Stanford, University of California–Santa Barbara, and Texas. To arrive at the percentage, I included in the denominator only department faculty members who had U.S. doctoral degrees.

[e] Percentage of department faculty with outside research support.

[f] Source: Supplement to Departmental Questionnaire.

[g] Missing data.

[h] Not evaluated by Goldberger, Maher, and Flattau (1995) because departments did not meet criteria for assessment. Supplement to Departmental Questionnaire not administered to the schools because they did not meet prior criteria.

[i] Percentage of department faculty with at least one graduate research assistant.

[j] Source: Departmental Questionnaire (see appendix F).

[k] In response to the following question: "Taking all the faculty members who do not have graduate research assistants, what would you say is the single most important reason?" (See Departmental Questionnaire, appendix F.)

l Technical support staff budgeted outside department; no numerical estimate given.

m Does not discriminate between courses and course sections.

n In addition to sabbaticals, faculty members may normally "buy out" a fraction of yearly courses through grants.

o Competitive research leaves for one term. No limit per individual. No more than 10 percent of faculty eligible for departmentally governed leaves in a given year.

p Refers to teaching a heavier course load to vacate subsequent terms from teaching duties.

q Number of undergraduates in the university. (The number of undergraduate majors in physics was not available for all departments. More important, however, such figures obscure the extent of undergraduate teaching in "service" and other general courses.

r Source: American Council on Education (2001).

s Source: Barron's Educational Series (2000).

t The maximum score for the Scholastic Assessment Test (SAT) (formerly the Scholastic Aptitude Test) and for the Graduate Record Exam (GRE) is 800 for each of their components.

u Includes only master's degree students; department no longer has a doctoral program.

v Excludes faculty salaries; refers exclusively to non-grant support.

w Amount depends on research program of specific faculty member.

x Presumably, faculty members rely on their research grants or unrestricted funds disbursed by the department or personally cover their travel costs.

y Estimated annual number of people outside the department who gave formal talks.

z In response to the following question: "As best as possible, indicate the extent to which teaching performance factors into promotion decisions in your department." Answer choices: a lot; somewhat; a little; practically not at all; not at all. (See Departmental Questionnaire, appendix F.)

many dimensions on which the data characterize these worlds, one would see clearly significant differences among them—objective differences that come to impinge in still more significant ways on individual careers and career outlooks, orientations, and satisfactions, as will be evident in the next chapters.

Among the notable changes over the ten-year period are differences in leave provisions from teaching, viewed by scientists as paramount for research. Leave provisions grew more strict at the elite and pluralist institutions, but somewhat more liberal at the communitarian institutions (where, in one case, leave provisions were introduced in the preceding ten years).

Annual federal support to the departments decreased appreciably (by $10 million) at the one elite institution for which there are comparative data, and gained only modestly in the one pluralist institution (by $3 million) and in the two communitarian institutions (by $1.7 million and $0.5 million, respectively) where comparative data is at hand. The overall departmental operating budget for the one elite institution where comparative data is available remained the same over the ten-year period, a remarkable no-growth rate. The annual departmental operating budget for the pluralist institution grew by $8 million, the greatest rate of overall change on this specific datum. One of the communitarian departments saw its annual operating budget grow measurably, by $2.2 million, but the other of the communitarian departments for which there are comparative data saw its annual operating budget grow just $10,000 in ten years.

The number of undergraduates increased at four of the universities whose physics departments are in this study. The increase was by as few as two thousand to as many as five thousand students, a demonstrable change in the demand for instruction. Average SAT scores also increased across the six institutions, on the order of fifty points. The two elite departments saw losses of approximately fifty in the total number of graduate students they enroll. The pluralist department saw a gain of sixty-five total graduate students; one communitarian department a gain of ten, another of five. The third communitarian department saw a loss of twenty graduate students enrolled in its program at a given time. Reimbursement for professional travel ceased in the two communitarian departments on which there are comparative data.

Finally, it became much more common over the ten-year period for department heads/chairs to explain a lack of graduate research assistants, used to support faculty research, by way of lack of funds. This was true particularly for the two elite departments and the one pluralist department.

An additional lens on departmental change comes by way of a component added to the questionnaires administered to heads/chairs in the longitudinal study. In this component, the heads/chairs were asked to evaluate their departments on twelve items that measure change in the conditions of work. These items run a range from operating resources, equipment and facilities, funding for graduate students and postdoctoral researchers, to the importance and difficulty level in obtaining external grants, the difficulty level of obtaining tenure and promotion to full professor, and departmental morale. Table 18 presents the findings.

Departments are seen by their heads/chairs to change in significant ways during this ten-year period, mirroring several of the comparative data presented in tables 16 and 17. In some cases, in some departments, these changes come at the expense of work conditions. In other instances, work conditions are enhanced.

In general, constraints on *overall operating resources* are seen by heads/chairs to grow more severe. *Equipment and research facilities* are seen to improve in four of the six departments, but become much weaker in one of the elite departments. Overall, *funding for graduate students* is stringent and seen to have become more severe, though less severe in one of the communitarian departments. Constraints on *funding post-doctoral researchers* are mixed across the departments as viewed by their heads/chairs. They are seen to grow more severe in the elite departments, but less severe or about the same in the pluralist and communitarian departments. Heads/chairs believe it is just as difficult or, in three cases, easier to *publish in the top physics journals*. The three heads/chairs who believe it is easier to publish in the top physics journals are those who claimed their departments' equipment and facilities were stronger compared to ten years ago, perhaps partly explaining this pattern.

Constraints on *faculty travel funds* are seen by the heads/chairs as "about the same" or more severe, or much more severe. The data provided in tables 17 and 18 on reimbursement for professional travel suggests that, even where conditions are perceived to be "about the same," they remain stringent and severe.

The importance of *obtaining externally funded grants* has grown, at least in the minds of department heads/chairs, and undoubtedly in the minds of faculty whose heads/chairs impress upon them this strong desire for outside support of research. But among the items on which the department heads/chairs were asked to indicate change, it is with regard to the *difficulty of obtaining externally funded grants* where their responses are most uniform.

TABLE 18. Administrator-reported change in departments

Item	Elite		Pluralist		Communitarian	
	School 1	School 2	School 1	School 1	School 2	School 3
1. Compared to 10 years ago, how would you describe the constraints on your department's *overall operating resources*?						
Much more severe	X					
More severe		X			X	X
About the same						
Less severe			X	X		
Much less severe						
2. Compared to 10 years ago, how would you characterize the *equipment and research facilities* of your department?						
Much stronger			X	X		
Stronger		X			X	
About the same						
Weaker						X
Much weaker	X					
3. Compared to 10 years ago, how would you characterize constraints on *funding for graduate students* in your department?						
Much more severe	X					
More severe		X	X		X	
About the same						
Less severe				X		X
Much less severe						

4. Compared to 10 years ago, how would you characterize constraints on *funding for postdocs* in your department?
   - Much more severe
   - More severe
   - About the same
   - Less severe
   - Much less severe

5. Compared to 10 years ago, how difficult is it for a typical faculty member to *publish in the top journals*?
   - Much more difficult
   - More difficult
   - Just as difficult
   - Easier
   - Much easier

6. Compared to 10 years ago, how would you characterize constraints on *faculty travel funds*?
   - Much more severe
   - More severe
   - About the same
   - Less severe
   - Much less severe

(continues)

TABLE 18. (continued)

	Elite		Pluralist		Communitarian		
Item	School 1	School 2	School 1	School 1	School 2	School 3
7. Compared to 10 years ago, how would you characterize the importance of faculty external *grants*?						
Much more important						
More important		X	X	X	X	X
As important	X					
Less important						
Much less important						
8. Compared to 10 years ago, how would you judge the competition for faculty external *grants*?						
Much more competitive	X	X	X	X		
More competitive					X	X
As competitive						
Less competitive						
Much less competitive						
9. Compared to 10 years ago, earning tenure in your department is . . . ?						
Much more difficult				X		
More difficult		X	X			
About as difficult	X				X	X
Less difficult						
Much less difficult						

10. Compared to 10 years ago, earning promotion to full professor in your department is . . . ?

Much more difficult				
More difficult	X		X	
About as difficult		X		X
Less difficult				X
Much less difficult				

11. Ten years ago, how would you have judged the morale of your department in terms of the ability to conduct and complete scientific work?

Very high	X			
High		X	X	
Average or fair				X
Low				X
Very low				

12. Compared to 10 years ago, how would you now describe the morale of your department in terms of the ability to conduct and complete scientific work?

Much higher than 10 years ago	X		X	
Higher than 10 years ago		X		
About the same as 10 years ago		X		X
Lower than 10 years ago			X	
Much lower than 10 years ago				X

TABLE 19. Federal spending for academic research

FEDERAL EXPENDITURES FOR ACADEMIC RESEARCH AND DEVELOPMENT (R&D), 1975–2001, FOR PHYSICS[a]

Year	Amount[b]	% Change
1975	449	—
1980	576	28.3
1985	750	30.2
1990	978	30.4
1995	1,009	3.2
2000	1,128	11.8
2001[c]	1,130	0.2

PERCENT FEDERAL DISTRIBUTION FOR ACADEMIC RESEARCH AND DEVELOPMENT (R&D), 1975–2001, FOR PHYSICS[a]

Year	%
1975	5.1
1980	5.3
1985	5.7
1990	5.2
1995	4.5
2000	4.0
2001[c]	3.8

(continues)

They agree that successful grantsmanship has grown substantially more competitive.

In three cases, *earning tenure* is viewed by heads/chairs as more difficult or much more difficult, and in the balance of the other three cases, about as difficult as ten years ago. The same balance of cases is true—although arrayed differently across departments—for *earning promotion to full professor*: it has grown more difficult, or is perceived to be as difficult as it was ten years ago.

Despite these conditions, the heads/chairs, in four cases, rated the *morale of their departments ten years ago* as "high," and in the balance of the two other cases, as "average or fair." In three cases, they *now rate the morale of their departments* as higher or much higher. The balance of the three other cases now rate the morale of their departments as "about the same as 10 years ago." This may be an accurate depiction of their department's collective

TABLE 19. *(continued)*

FEDERAL EXPENDITURES FOR RESEARCH EQUIPMENT AT ACADEMIC INSTITUTIONS, 1985–2001, FOR PHYSICS[a]

Year	Amount[b]	% Change
1985	97	—
1990	106	9.3
1995	117	10.4
2000	103	−12.0
2001[c]	103	0.0

FEDERAL EXPENDITURES FOR BASIC RESEARCH, 1955–2000[a]

Year	Amount[b]	% Change
1955	589	—
1960	1,534	160.4
1970	4,551	196.7
1980	5,366	18.0
1990	7,962	48.4
2000[c]	12,992	63.2

Notes:
[a] Source: National Science Board (2004).
[b] In millions and in 1996 constant dollars.
[c] Latest year for which non-estimated data are available.

sentiments. It may also partly reflect a desire by heads/chairs to view the departments they oversee in a favorable light.

The relative uniformity of responses concerning the substantially increased difficulty of obtaining external funding, along with the result that fewer graduate research assistantships exist because of a lack of funding to create them, points to perhaps one of the most critical trends in academic physics (and in many other fields of the academic profession). As it turns out, it is a trend that will resurface nearly incessantly in the career accounts presented and analyzed in the subsequent chapters. This trend broadly encompasses the net decline in federal support for physics specifically and for academic research and higher education generally.

Trends in U.S. federal support of physics, through research and development (R&D) and equipment, are presented in table 19. The data reveal highly significant changes over time. What is more, the changes are particularly acute since roughly the early 1990s. The data show that federal support for physics becomes even more stringent at the turn of the century. Federal research and development expenditures, while increasing over time, have decreased markedly in their rate of change over time. This is reflected in the percentage of federally distributed monies for academic R&D to physics, which shows an overall steady decline since 1975, and a particularly steep plummet from 1995 to 2001, the last year for which data are available. At the same time, federal expenditures for research equipment in physics at academic institutions saw a substantially sharp decrease from the mid-1990s on into the next century.

Federal expenditures for basic research across fields, presented in the final panel of data, show that heightened fiscal constraint over time characterizes not just physics but a wide research spectrum. The percentage changes reflect gains, but the percentage increases are vastly different from thirty years ago, or since roughly 1970 and the preceding decades. The percentage increases declined especially in the 1970s, and grew modestly in the 1980s and 1990s when compared to previous decades. These data convey ever so strongly that the academic conditions for work in physics have changed rather dramatically, and quite notably in the last decade of the twentieth century. This constitutes a period when a sizeable fraction of the scientists in this study are attempting to cultivate their careers. But they are attempting to cultivate their careers in different phases: some in the relative funding heyday of the 1960s, others under comparatively greater financial pressure in the 1980s and 1990s.

Other research has revealed trends that are consistent with what has been found here. The trends appear to cut across academic fields generally, with only handfuls of exceptions to the general pattern. For instance, Finkelstein, Seal, and Schuster (1998) found through a national faculty survey that 37.7 percent of all faculty reported worsened conditions for obtaining external funding. Only 18.4 percent of faculty believed that such conditions had improved over time. Over 50 percent reported pressures to increase their workloads in the midst of constrained resources (Finkelstein, Seal, and Schuster 1998, 95).

Overall, the conditions above speak of constraint and growing stringency. When viewed in conjunction with the conditions of practice presented in tables 16 and 17, the academic worlds remain significantly differentiated from one another. These conditions form the backdrop against which to begin to assess the unfolding of careers in their institutional environments.

In the chapters that follow, I turn to the scientists and their careers. I begin with the youngest of them, those who have passed from early to mid-career, and will then work my way over the successive chapters through the other, progressively more senior cohorts in order to see what scientists learn about their lives in science.

CHAPTER TWO

Early- to Mid-Career Passages

In this and the next two chapters, I concern myself with how each of the three cohorts of scientists, beginning with the youngest and proceeding to the middle and finally the eldest, progressively passed through their respective sets of career phases. I do so with a theoretic objective: to examine how scientists age in relation to their work. Responsive to the more general sociological concerns that situate the study of professional careers in science over time, as discussed in the introduction, each of these chapters will emphasize the biographical ways in which scientists account for the turning points and transitions that socially situate their careers in the institutions where the scientists have worked. In so doing, I will call attention to how the professional life course in science may be differentiated, and with what consequences to careers, to the individuals attempting to develop them, and to the institutions that socially shape them.

PROFESSIONAL PROFILE

It is useful to provide a professional context for each of the cohorts, beginning with the early-to-mid-career scientists. Such a context will help inform the accounts of the scientists presented subsequently. If one is concerned with how scientists age in relation to work, then aspects of age itself along with work, especially having to do with the roles most valued by the occupational group in question, are central pieces of data that provide a professional context of the scientists. This is the function of table 20.

It was possible to earlier glimpse (in table 13) the ages and lengths of professorial experience of the scientists for each cohort. Table 20 breaks down the figures further by the three academic contexts.

In table 20 are listed the age ranges, average ages, and range of years of experience as a professor at the time of the first interview, followed by comparable data for when the ten-year follow-up interviews were completed.

TABLE 20. Cohort characteristics: The early- to mid-career cohort[a]

Characteristics	Elites	Pluralists	Communi-tarians	Overall Average
AGE & EXPERIENCE[b]				
Age Range @ 1st Int.	32–39	35–41	33–45	33.3–42.0
Avg. Age @ 1st Int.	34.4	37.0	39.0	37.0
Age Range @ 10Yr. Fw-Up	42–49	45–51	43–55	43.3–52.0
Avg. Age @ 10Yr. Fw-Up	44.4	47.0	49.0	47.0
Range Yrs. Exp. @ 1st Int.	2–6	4–7	3–9	3.0–7.3
Avg. No. Yrs. Exp. @ 1st Int.	4.1	5.2	6.0	5.0
Range Yrs. Exp. @ 10Yr. Fw-Up	12–16	14–17	13–19	13.0–17.3
Avg. No. Yrs. Exp. @ 10Yr. Fw-Up	14.1	15.2	16.0	15.0
JOURNAL PUBLICATIONS: QUANTITY[c,d]				
Maximum @ 1st Int.	66	37	32	45.0
Minimum @ 1st Int.	12	13	10	11.6
Mean @ 1st Int.	29.3	23.9	21.7	25.0
Maximum @ 10Yr. Fw-Up	124	68	94	95.3
Minimum @ 10Yr. Fw-Up	41	27	17	28.3
Mean @ 10Yr. Fw-Up	73.0	44.0	50.0	56.0
JOURNAL PUBLICATIONS: QUALITY[e,f]				
≥80% published work/1st Int.	75.0	71.4	0.0	48.8
≥70% published work/1st Int.	88.0	71.4	14.3	58.0
≥60% published work/1st Int.	88.0	86.0	43.0	72.3
≥80% published work/10YrFwUp	75.0	67.0	0.0	47.3
≥70% published work/10YrFwUp	88.0	67.0	14.3	56.4
≥60% published work/10YrFwUp	88.0	67.0	14.3	56.4
ROLE PERFORMANCE				
Avg. No. Papers @ 1st Job[g]	20.0	12.0	11.0	14.3
Avg. No. Papers @ Tenure[h]	46.0	26.0	24.0	32.0
Avg. No. Paper @ Full Prof.[i]	54.0	39.3	38.0	44.0
Avg. Time to Tenure (in years)[j]	7.0	5.2	5.3	5.8
Avg. Time to Full Prof. (in years)[k]	5.0	4.8	6.0	5.3

(continues)

TABLE 20. (continued)

Notes:
[a] Source: Each scientist's curriculum vitae.
[b] Elites: N = 8; Pluralists: N = 7; Communitarians: N = 7. Includes all members of the foundational study sample.
[c] The number of publications for each scientist includes all published journal articles, scientific journals being the primary medium through which scientists disseminate their research. The number excludes the following: books; textbooks; book chapters; edited volumes; conference proceedings; invited and contributed papers; book reviews; encyclopedia, world book, and yearbook entries; and articles listed on the individual's curriculum vitae as "submitted," "in press," "accepted for publication," "in preparation," and so on. If the same journal article was published multiple times (in different venues), it is counted only once.
[d] For foundational study: Elites: N = 8; Pluralists: N = 7; Communitarians N = 7. For longitudinal study: Elites: N = 8; Pluralists: N = 6; Communitarians: N = 7.
[e] Percentages of scientists' publications appearing in "major physics journals," those journals to which the physics community assigns greatest value. These include, in alphabetical order, *Astronomical Journal*, *Astrophysical Journal* (including *Supplement Series*), *Astrophysical Letters and Communications*, *Europhysics Letters*, *Geophysical Research Letters*, *Icarus*, *International Journal of Modern Physics* (including series A, B, C, D, E), *Journal de Physique* (including series I, II, III, but excluding IV), *Journal de Physique Lettres* (now incorporated with *Europhysics Letters*), *Journal of Chemical Physics*, *Journal of Geophysical Research* (including series A, B, C, D, E), *Journal of Mathematical Physics*, *Journal of Physics* (including series A, B, C, D), *Lettre al Nuovo Cimento* (now incorporated with *Europhysics Letters*), *Nature*, *Nuclear Physics* (including series A, B), *Physical Review* (including series A, B, C, D, E, L), *Physics Letters* (including series A, B), *Physics of Fluids*, *Review of Modern Physics*, *Science*, and *Solid State Communications*.
[f] For foundational study: Elites: N = 8; Pluralists: N = 7; Communitarians N = 7. For longitudinal study: Elites: N = 8; Pluralists: N = 6; Communitarians: N = 7.
[g] Average number of journal articles published by the time scientists were appointed to their first job. Other publications excluded. "First job" is defined as appointment as assistant professor or visiting assistant professor. Elites: N = 8; Pluralists: N = 7; Communitarians: N = 7.
[h] Average number of journal articles published by the time scientists earned tenure. Other publications excluded. Elites: N = 8; Pluralists: N = 6; Communitarians: N = 7. Cases do not agree with total cohort numbers because of unavailable data. Scientists who began their careers outside of academic science (e.g., as industrial scientists) are excluded.
[i] Average number of journal articles published by the time scientists were promoted to full professor. Other publications excluded. Elites: N = 7; Pluralists: N = 4; Communitarians: N = 6. Cases do not agree with total cohort numbers because of unavailable data or because scientists are not full professors. Scientists who began their careers outside of academic science (e.g., as industrial scientists) are excluded.
[j] Average time in years it took scientists to receive tenure. Elites: N = 8; Pluralists: N = 6; Communitarians: N = 7. Cases do not agree with total cohort numbers because of unavailable data. Scientists who began their careers outside of academic science (e.g., as industrial scientists) are excluded.
[k] Average time in years it took scientists to earn promotion to full professor. Elites: N = 6; Pluralists: N = 4; Communitarians: N = 6. Cases do not agree with total cohort numbers because of unavailable data or because scientists are not full professors. Scientists who began their careers outside of academic science (e.g., as industrial scientists) are excluded.

The scientists who have passed from early to mid-career now average in age from their mid- to late forties, pluralists slightly older on average than elites, and communitarians older still on average than either elites or pluralists. When I first interviewed this cohort of scientists, they had worked as university professors an overall average of 5 years. This range of experience extended to as few as 3 years and as many 7.3, by no means numerous in the context of a full academic career. Ten years later, at the time of the follow-up interviews, age and experience are considerably transformed. These scientists have now been university professors for an overall average of 15 years, the overall average range extending from 13 to 17.3 years.

As table 12 established, the scientists who passed from early to mid-career advanced significantly in professorial rank. The significance lies not only in the ranks attained but also in the difficulty and momentousness of achieving those ranks, most especially the rank of associate professor from assistant professor and the (usually) concurrent awarding of permanent academic tenure. Advancement to the rank of full professor, while also arguably difficult and momentous, is typically viewed and experienced by scientists (and other academics) as a different type of status passage: it is one in which one's job (and professional future) is not on the line, as is the case with advancement from assistant to associate professor, but rather one where evidence is presented that a scientist-scholar has produced a body of work to merit attainment of the pinnacle professorial rank in the institutional academic career. Of the twenty-one members of the early- to mid-career cohort, seventeen are now full professors, having been either assistant or associate professors when I first interviewed them. The remaining four professors are associates, three of whom were assistants when I first interviewed them. One member of this cohort was and remains an associate professor.

That which currently enables scientists to advance in rank today is publication productivity, the component of the academic role that has assumed a decided prominence in the organization of academic life across nearly all types of academic institutions, from the research university to the comprehensive university and the liberal arts college, and even in more muted (but noteworthy) ways to the community college, attesting to its charismatic-like power in permeating the ever-wide array of academic institutions (Blackburn and Lawrence 1995; Finkelstein, Seal, and Schuster 1998). Table 20 presents data on the quantity and quality of publication productivity of scientists for each of the points in time I interviewed them.

At the time of the first interviews, elites out-produced pluralists who in turn out- produced communitarians, reflecting a pattern of stratification that mirrors the prestige of the departments and universities in which the scientists work. The patterns of productivity change, however, over time. Ten years later, elites continued to produce the greatest volume of work, having published a range of between 41 and 124 articles, or an average of 73 articles. A striking pattern develops in comparison with the pluralist and communitarian scientists, though, wherein the publication productivity of the latter exceeds the former. On average, communitarians have written 50 articles compared with the pluralist average of 44, and the maximum number of articles produced by a member of the communitarian academic world exceeds the maximum produced by a member of the pluralist world by a substantial margin (94 to 68).

One might have expected the publication productivity of pluralists to have advanced more competitively with elites, so as to assume a more intermediate position comparable to that found among the base-year study publication patterns. One possible explanation for this aberration may be the presence of highly productive communitarian scientists, who elevate both the communitarian maximum and the communitarian average significantly above their pluralist cohort counterparts. Yet the pluralist cohort contains able scientists who are capable of producing in ways akin if not identical to their elite peers. As it turns out, the piece to this puzzle—what in particular has happened to the pluralist cohort to dampen their comparative productivity—will be found later in the interview accounts provided by the scientists.

In addition to publication quantity, the quality of publication provides an additional measure of this most central aspect of academic role performance. Here, too, stratification is evinced in performance across the three contexts of academe. Elites publish at comparatively high-quality thresholds. What is more, their rates of publishing at these thresholds were unchanged over the interval of time spanning the two rounds of interviews. Seventy-five percent of elites in this cohort published 80 percent or more of their work in major scientific journals. Eighty-eight percent of elites published at the lower thresholds, where 60 percent or more of their work appears in the major scientific journals.

Publication quality among pluralists was roughly comparable to elites at the time of the first interviews, but dropped by the time of the second interviews: 67 percent of pluralists published a majority of their work in the major journals. The patterns are different still for communitarians. When first interviewed, this cohort of communitarians did not publish a majority of their work in the major scientific journals. The percentages fell over the ensuing ten years such that only 14.3 percent of this communitarian cohort published 60 percent or more of their work in the major journals.

Variation in publication quantity and quality convey different norms of role performance governing professional life in these different types of universities. Institutional expectations of role performance differ. But professional outcomes, that is, advancement in rank, are nearly the same across the three types of academic worlds, a point underscored by the final panel of data on role performance presented in table 20. All of these scientists achieved tenure, albeit at slightly different rates (longest in length among elites), and numerous members of this cohort have advanced to the rank of full professor (again at slightly differing rates across contexts). Despite these similarities in professional outcome, the institutional expectations for role performance have varied considerably, particularly between the elite and non-elite subsets of the scientists.

This cohort of elites had published an average of 20 papers by the time they were offered their *first* professorial appointment, a rather remarkable threshold for entry into academe, and one that will prove even more stunning when later compared to older cohorts of scientists. Pluralists and communitarians of this cohort had published an average of 12 and 11 articles, respectively, in order to enter academe; while substantially different from elites, this, too, may be viewed as a remarkable achievement as well as a remarkable expectation, particularly when later compared to the expectations and achievements of older cohorts of scientists.

Young elites had published an average of 46 papers at the time they earned tenure, compared with 26 and 24 papers, respectively, for pluralists and communitarians. Though the figures vary especially between elite and non-elite subsets, the figures for each of the contexts are substantial in magnitude. The same observations hold for the numbers of papers scientists had on average published by the time they were promoted to the rank of full professor (for those who had attained this rank by the time of the second interviews): 54 for elites, 39.3 for pluralists, 38 for communitarians.

While data on publication quantity and quality are important means by which to differentiate academics, one must also recognize that these are imperfect measures of performance. Publications themselves in top-ranked journals may, or may not be, top-notch. Similarly, good ideas at times do not see the light of day in top-ranked journals; they appear elsewhere. By the same token, some academics publish comparatively less in quantity but, given the quality and impact of what they have published, enjoy a reputation that is comparable to more prolific researchers. Likewise, some academics are "mass producers," but given comparatively less recognition paid to their work, have found less renown than "perfectionists"—those who have published less but of higher quality (Cole and Cole 1967). While it behooves one to note the imperfections of these measures, the measures nevertheless entail significant consequences. The measures are used by the academic community, by departments and universities, and thus by individual academics to draw distinctions about performance. For example, quantity and quality of publication have become the penultimate devices used by faculties, albeit in varying combinations, to determine tenure and promotion through the academic ranks in universities.

These data provide an outline of change. They convey the extent to which the scientists in this cohort have aged, along with the number of years of experience they have gained as professors. The data capture the scientists' activity in the central academic role of publication and track how this has changed over time in both quantity and quality of productivity. It is therefore possible to observe what it has taken, in terms of scientific output, in order for scientists to climb the academic career ladder. But to understand what this career passage *means*, one must go to the scientists and their subjective accounts of their careers. In order to evaluate the ways in which the scientists have *changed* in the meanings they ascribe to their careers, one must go to the scientists more than once, so as to capture differences over time. Thus, to examine how scientists age in relation to their work, one needs to understand the career meanings *from which they have come*. It is necessary to understand how scientists understood themselves at a previous point in time, in their early careers, in order to glean how their career outlooks and orientations have, or have not, changed from their present outlooks at mid-career.

EARLY-CAREER PATTERNS

A series of generalizations, which may be made about the early-career patterns of this cohort of scientists, will help establish a base on which longitudinal inquiry can build. The goals of these generalizations are fundamentally twofold. The first is to provide a parsimonious overview of the findings about this cohort as studied in the foundational work so that readers may be aware of the subjective views of careers from which these scientists have been followed. The overview shall be parsimonious in order to provide a sufficient accounting of these scientists' pasts, such that one may more completely understand them in the present, without reproducing the bulk of the prior work. The second goal is to establish a base of understanding on which one may build subsequent inquiry into these careers. The present study is of course not wholly independent of the former; the scientists link the two studies together. Hence one of the chief aims is the development of *cumulative knowledge*. In this work, a unique opportunity exists to examine careers diachronically. One may thus come to view the academic profession and academic careers in terms of cumulative knowledge about experience and interpretation of work by its practitioners through time.

Generalizations may be made concerning seven dimensions of careers that arose from the foundational study and which help to distinguish career patterns across the three prototypes of academic worlds. The seven dimensions of careers include: career focus, professional aspirations, recognition sought, orientation to work, work/family focus, attribution of place, and overall satisfaction (Hermanowicz 1998, 2002). These dimensions of careers along with the substantive findings pertaining to them, which arose from the coding of data in the foundational study, are presented in table 21 for each of the three academic worlds.

Elites, pluralists, and communitarians focused centrally on scientific research in their early careers, the expected fruits of which would translate into promotion and tenure. This was true across the academic contexts and across all individuals, even as some institutions espouse and emphasize and as some individuals value and identify with their roles as teachers. The scientists were socialized and trained, while as graduate students, often as postdoctoral researchers, and as new faculty members, that achievement in research above all else would work most readily on behalf of their professional futures and job security in academe. This is not to say that scientists valued job rewards above the satisfaction they derived from scientific research in and of itself; it is precisely this satisfaction and its anticipation that drew them to an academic career. Once in their organizational environments, the scientists became aware, more so than at any previous time, of the realities and expectations of their role performance. These realities were felt especially by this cohort of scientists, whose careers were conditioned by decidedly heightened success norms, as data on role performance presented in the next two chapters will demonstrate.

It was in these first years as assistant professors or early associate professors when major changes in professional outlook and orientation to work began to take root. A consequence of these incipient shifts is observed in the scientists' professional aspirations at this time. Aspirations intensi-

TABLE 21. Early career patterns of scientists

Career Dimensions	Elites	Pluralists	Communitarians
CAREER FOCUS			
In Early Career	Research	Research	Research
PROFESSIONAL ASPIRATIONS			
In Early Career	Intensify	Rescaled	Diminish
RECOGNITION SOUGHT			
In Early Career	Great	Great	Great
ORIENTATION TO WORK			
In Early Career	Moral	Moral	Moral
WORK/FAMILY FOCUS			
In Early Career	Work	Work	Work
ATTRIBUTION OF PLACE			
In Early Career	"Burden"	"Happy Medium"	"Stymieing"
OVERALL SATISFACTION			
In Early Career	Medium	High	Low

fied among elites, whereas among pluralists aspirations were rescaled in ways that were more realizable within the structure of constraint and opportunity of their work environment. Aspirations among communitarians were substantially diminished in their early careers, owing to a significantly more stringent structure of opportunity and constraint upon their research careers.

As the foci of careers began to change, particularly among pluralists and even more so among communitarians, often beginning in the first three years as assistant professors, the recognition that scientists sought from the professional community of scientists began to evolve in systematically distinct ways. Recognition endures as a chief property of scientific careers because of its function of providing socially certified testimony, by those competent to judge contributions, that a scientist has indeed satisfied the institutional goal of science by extending knowledge. The level of recognition sought by scientists in each of the three academic contexts may be characterized as great, or substantial—nearly all scientists, especially in their early careers, entertained the idea, or perhaps the myth, of their own role in a great discovery. But the accounts from these early-career phases, together with the ways in which aspirations began to differentially unfold, suggested that the recognition sought by scientists would itself change greatly in subsequent phases of the career.

Beginning their careers with a focus on research, elites, pluralists, and communitarians adopted a moral orientation to their work. That is, science and scientific research were viewed by the scientists as ends in themselves. This orientation is compatible with the institutional goals of science: scientists conduct research because their role morally mandates this activity. By extension, were they not to engage in this activity faithfully, and if they were properly socialized for this role as members of the academic profession, then the occasion would call for a scientist to interpret meaningfully for him or herself and for others why there had been a shift in, or perhaps even a withdrawal from, the performance of this role.

In ways consistent with their focus on research and the search for recognition, elites, pluralists, and communitarians adopted a focus on work, independent of nearly all other concerns, including family. This was true even in this period when early career often coincided with family onset: the birth and rearing of children. The rescaling of professional aspirations, however, along with anticipated modulation in the recognition scientists sought, suggested that this might change in all or in selected academic contexts over ensuing phases of the scientists' careers.

These patterns led to overall characterizations of work-place and the person. I argued in the foundational work that the elite world is typically viewed by elites as a "burden" because institutional mandates for achievement and the internalization of substantial performance norms demand sustained effort. Succeeding in the elite world by attempting to satisfy these institutional expectations was typically interpreted by elites as a type of obligation: in order to be viewed by peers as bona fide members, they believe they must deliver on their socially ascribed promise as scientific scholars. For these reasons, satisfaction among elites was best characterized as "medium." While their institutional location provided an objective measure of prestige, institutional expectations were such as to continually prompt sustained performance at a high level, of which the publication quantity and quality measures discussed previously are testimony. Many elites never felt fully satisfied in light of these demands.

The communitarian world was viewed by communitarians as "stymieing," particularly because of the adjustments they saw themselves having to make in their professional aspirations, conditioned and constrained as they were by their local work environment. Consequently, satisfaction among communitarians could best be described as "low," both in real terms and in comparison with how peers in the other contexts of science viewed their satisfaction.

The pluralist world was viewed by pluralists as a "happy medium," largely because this type of academic world allowed its members to pursue their work without sacrificing at great magnitudes their professional aspirations. At the same time, the pluralist world did not impose the heightened expectations characteristically found among elites. These conditions resulted in relatively high overall levels of satisfaction among pluralists.

Having developed these various outlooks and orientations on their work as their careers began, the scientists are now at middle phases of their careers. Their employment in academe is secure, although the paths to that security were sometimes difficult and nearly always anxiety filled, as their accounts will reveal. Many have ascended to the pinnacle academic rank of full professor, conferring additional measures of security and attainment, although in some cases in some academic worlds this may also confer unanticipated changes to individuals and to their careers. To be certain, as their ages alone begin to mark, this cohort of scientists now commands considerably more years of experience as scientific researchers, as scholar-teachers, and, as an indication of some of the changed roles scientists assume, administrators in the organization of science and academia. How do careers unfold across the settings of science, with what possible variations, and with what consequences to individuals and their identification, commitment, motivation, and the satisfaction that they bring to and find in their work? The remainder of the chapter attempts to answer these questions by turning attention to the middle phases of these scientists' careers.

ELITES

> The dream is to discover some fantastic new effect that knocks the socks off my friends and colleagues, that knocks the socks off the community, so that when I walk down the corridor, the young students

know me and say, "There goes [Silverman], he invented the [Silverman] effect." That's what I want; I want my effect. I want to be the first person to predict such and such an event and for it to be . . . I can even smell what it's like already. It has to be something which once you think about it, is very reasonable. Very surprising at first sight, but at second sight, yes, of course, that's how it had to be. I want one of those. I want my Josephson effect, my fractional quantum Hall effect. (23F)[1]

Such was the future envisioned by a scientist who, given the pseudonym "Geoff Silverman" when interviewed in the foundational study, was a thirty-four–year-old recently promoted associate professor doing science at one of the premier research universities in the United States. Like nearly all of his elite contemporaries, he was intensely clear about his achievements, those made and those planned for the future; about his role, real and imagined, in science; and about his talents, demonstrated and hoped-for, as a researcher, scholar, and teacher—all a testament to the high level of recognition he and the scientists of his cohort sought for superior scientific achievement. A celebrated scientific role was so clearly envisioned, and so well defined, as to be defined out of the reach of attainability, akin to what the social psychologist Daniel Levinson described as "the dream"—a term that this scientist himself uses to describe his anticipated future—and its resulting "tyranny," or overweening power that must be reconciled, purportedly in mid-life, according to Levinson, in order for individuals to exist in a state where their expectations are in greater alignment with capabilities and opportunities for success (Levinson 1978). Through the early phases of his career, the dream of great attainment carried immense power and thus immense importance in defining his vision of himself and an academic career.

> It's true that every physicist has a check or not beside his name at the end. No matter how good you are, if you don't get the Nobel Prize, it really matters. It matters immensely. It's almost as important as sex differences. There are the Nobel Prize winners, and there are the others. It's probably not healthy, but my picture of the physics community is absolutely that. There is this branch of people, this narrow band of people who sit at the high table, and everyone else is an also-ran. . . . We all dream about walking into the travel agency and asking for our tickets to Stockholm. We even joke about it. We talk about those colleagues of ours who get twitchy every time the phone rings in October,

because you know that's when the phone call comes—something we openly discuss, most of us. I probably think that anybody who says they don't want the Nobel Prize is probably lying. We all want that ultimate seal of approval. I certainly do. It will give you the opportunity to thumb your nose at the few people that you've got tangled up with over the years. (23F)

Ten years later, self and role in science are newly viewed, the dream of great attainment more closely reconciled with capability and opportunity to see it fulfilled.

It clearly must have been a vain wish . . . I was more "effect driven" then. I remember being interviewed for a job at Oxford University, and they asked me something like, "What do you want to do?" and I said, "I want to knock the socks off all the other physicists in the world with such and such." I remember them just flinching and recoiling. I shouldn't say such a thing. So, how interesting . . . I know exactly what you're saying, I was speaking of the Josephson effect. I was obsessed with it at the time, I had my own set of versions of it. I never even published that piece of work. I have a hundred-page manuscript sitting in my computer that I haven't published. But I was completely smitten with those feelings. . . . That has definitely slipped off the radar screen. . . . I would say that my previous remarks were unrealistic, . . . I see that the opportunities for such events are so rare as to be not part of the normal course of science. . . . [My aspirations] have sort of mellowed. They are more, I don't like the word, but they're sort of holistic. I want my entire career to be successful, so my measure has changed. I see merit in sustained brilliance rather than unique episodes of brilliance. (23L)

As self and role in science evolve, so do markers—such as perceptions of the pinnacle scientific prize—that situate and help define scientists. Those markers, and in particular this prize, fade in the intense prominence they earlier played in morally constructing scientific careers, but they do not fade into oblivion—a point as true for the elite members of this cohort, as for the elder cohorts.

In all honestly, I don't think about it daily, but it comes across my mind every now and again, and I think to myself, "You're lucky to be in the building," and I really do think that most of the time. I know many people with comparable skills or higher skills whose life chances didn't put them in this seat. So, mostly, I just say to myself, "I'm paid to play in the sandpit that I love, and anything beyond that is really a bonus." (23L)

In these accounts—and in others below that will help bring the image of scientists moving from one point of their careers to another into even greater resolution—one is able to see several of the defining qualities that mark elites' passage from early into middle phases of their careers. Most prominent among these are what may be taken as twin core characteristics that situate this passage: *stabilization* and *rededication.* In stabilizing, individuals and careers are understood and experienced by scientists as more secure. This is sanctioned institutionally by promotion in academic ranks and by the conferral of tenure. Through rededication, individuals and careers are pledged once again, ritually through promotion and tenure, to the goals of science—to extend certified knowledge through research. Especially in the case of young elites, when status transitions proceed affirmatively, they operate not only to celebrate and reward the achievements that have led up to them, but also to stimulate sustained work. Empirically, the data suggest that these occur as twin processes. Without one, the essence of what this passage means to the scientists experiencing it is lost. They are core processes because, in the absence of them both, it is impossible to ascertain the empirical reality that comes to characterize this transition for elites.

In [these] ten years, I think you go from being a junior person trying to struggle to being more a senior statesman in the department. From the point of view of guiding the direction of this department, I have a lot of influence right now, whereas before I wouldn't even have tried to make direction changes in the department that regard hiring, the curriculum, things of that sort. . . . Before I would have been more of a team player in a sense of saying, "You're the boss. You say what my job is. I teach that course." Now I say, "This is what I think students here need to have," and people say, "Okay, we'll give you time to develop that course." Parallel to that, the project that I had started ten years ago—which was absolutely a risky project for me to jump in on at the time because it was a change in direction—really bore tremendous fruit in the last few years. It was at the level of the front page of the *New York Times* a couple of times. We have been traveling all over the world giving talks on the results of the project. It has been so much fun to ride a project that gets a lot of attention. (22L)

The intense professional aspirations of elites' early careers, along with the high level of recognition they sought especially as assistant professors, stabilize. If this occurred as an isolated developmental pattern, aspiration and the drive for recognition might be seen to wane. But, for nearly all elites, this does not occur, owing to this co-present developmental pattern of rededication. That the object of this rededication is solely *scientific research* is itself a finding, since a variety of roles customarily composes the academic career. As a scientist observed in a statement voiced ubiquitously among his elite contemporaries, "It's really research that has dominated my intellectual thinking and my passion about university work" (23L). To these elite scientists, the fact is understood more nearly as a truism: "The biggest joy, of course, is just researching and results. . . . I remember each day what we were thinking, what we were doing, what we were thinking when we got a particular insight. Those, of course, are the real high points" (4L).

Rededication, however, implies recognition of personal change. This change, centered on the conferral of tenure and on promotion, particularly to the rank of associate professor, is substantial. It is a change customarily anticipated years in advance. Particularly among members of this age cohort, the anticipation of tenure prompts tremendous publication productivity in order to achieve a successful outcome. The security brought about by promotion and tenure extends beyond the strictly institutional to the more personal, helping to establish in these scientists a secure feeling about themselves *as scientists*, expressed in newfound self-confidence as well as independence in attitude about nearly all aspects of their roles.

It will be increasingly apparent that this theme of rededication is most prominent among the elite subset of the early- to mid-career cohort of scientists. It is not a core attribute that defines this passage by pluralists or communitarians. This is true perhaps because of the expectations governing the role performance of elites, so amply conveyed by the productivity norms in table 20. In the transition from assistant to associate professor, much is at stake, and much is expected. The numbers of table 20 tell one side of the story; the scientists' accounts tell another.

> I have a pronounced sense of insecurity. I don't mean I want people to like me or have lots of friends. That's not the issue at all. This issue is genuine security, that is, having food on the table and a house to live in.

I always had a fear of being hungry and without housing.... So I've always felt particularly vulnerable, and, of course, some people will argue that that's absurd: if you have a PhD in theoretical physics, the minimum you could do is walk into anywhere in the world and teach in a high school. And that's probably true, but I don't have that job. I haven't got the offer in writing. I must confess I have felt tremendous insecurity, and the tenure issue was one that lived with me daily for six years. I saw my life only in terms of existence in the length of time between now and when my tenure package would be complete.... This issue completely dominated my mood, and my decisions, and my choices. For example, I wasn't prepared to have a child until I had tenure. I was very cautious with money. I would forgo holidays and enjoyment, and I worked day and night. Lurking behind it was this enormous need to accomplish as much as I could before this curtain came down. And this is incredibly stressful. And what is strange, rather like with childbirth, where there is some sort of release of biochemical agent which helps the mother forget the pain, I think I've really forgotten the anxiety. I can't really tell you quite what it was like. I'm a polite person and generally good-natured and cheerful and optimistic, and yet I would call up complete strangers, for example, editors of journals and argue with them, yelling and screaming on the telephone. I would scream and shout at baggage attendants at airports if my baggage didn't arrive. I was just a really unpleasant person, all driven by this fear that at the end I would be rejected and not able to have my salary and my house and my food. It's really at that primitive level.... I never got a signal from anybody that there was ever any glimmer of doubt. And despite all that, there was this enormous fear and stress, and really terror.... I think I had a reputation... of being a remarkably friendly and positive and cheerful soul. Nevertheless, inside I was really crazy.... The only people who understand are people who have been through it, and not all of those, because some people go through it in a completely different way. But I really think it's one of those passages in life which can only be understood by people who at least have been through that. It's the same as many other things in life. Probably going to Vietnam is something you can only understand really if you went. And it's the same in going through tenure. It's a singularly stressful event. (23F)

Ten years later, at mid-career, the prominence of this achievement remains keen: "Never does a day go by when I don't marvel on the notion of having tenure" (23L). For elites, however, the achievement does not change their fundamental orientation to work. Their orientation remains moral, their aspirations remain driven scientifically as their careers remain focused on research.

> I drive myself as hard as I ever did. I don't feel any sense of slowing down. There are subtle style shifts, no sense of slowing down. . . . I'm much more mild-mannered. . . . I don't think I feel stress anymore. I now have children . . . they are much more important to me than I ever imagined, . . . I think I was writing to impress other people and other ordinances, whereas now I'm writing to satisfy myself. . . . It's a very clear [shift], I am writing for me. It's almost like my own personal diary of science. . . . When I [now] think about my career, [this] is what I see: it's a sustained, growing, and substantial body of work. . . . I get a constant sense of pride from that . . . [there is] steady productivity and steady funding and steady success with students who get jobs. (23L)

> I don't feel less pressure. I feel more relaxed about my future, getting a secure position. But I feel enormous demands on my time.
> INTERVIEWER: Do you feel you're working harder, less hard, or about the same as ten years ago?
> SCIENTIST: About the same.
> INTERVIEWER: Do you think you're more ambitious now or less than you were ten years ago?
> SCIENTIST: About the same. (9L)

In the ten-year interim, one of the scientists of the youngest cohort rose from an assistant professor to holding an endowed professorial chair, a recognition for a line of superior scientific achievements in the area of theoretical physics. He put the substance of the evolution in his outlook in these terms:

> I have always been very engaged in my research. . . . Ten years ago, I had some outlook, and now I have an even brighter outlook. But this brighter outlook is simply because of progress of our research. . . . And it looks pretty exciting, actually. Even more exciting than what we knew ten years ago. That's all just a continuation. . . . I would say I'm more excited by my career and the research . . . [and I would say I am now]

more ambitious . . . I see the bigger picture . . . I see the bigger picture for my work, and that makes me more ambitious. (5L)

Short of rededication, the stabilized career could ease. Short of stabilization, the rededicated career could proceed haphazardly undirected. The account shows how, in combination, the elements of stabilization and of rededication produce an intensified identification with science. In this light, the scientific reward system operates in functionally intended ways by recognizing merit that in turn stimulates sustained commitment to scientific productivity.

The stabilization and rededication of careers by elite scientists in this set of career phases produces another behavioral manifestation, again more prominent among elites than any other group: satisfaction. Satisfaction with the career appears greatest among the elites of this cohort. Indeed, as the evidence will gradually come to show, this subset of scientists—elites of the early- to mid-career cohort—may well be at points in their careers when they are the most content compared with other types of scientists and to the other cohorts. Universally, when asked when in their careers they were the most satisfied, they identify present points in time.

I don't see what would be better than this right now. I want to quickly retire, so that it doesn't get worse. I can see it sliding negatively really quickly. I want to keep deflecting those possibilities. You have to keep pouring energy into it. If you stop pouring energy into it, it absolutely will go downhill instantly. (22L)

This is occasionally accompanied by identification of just one other point in time when they felt most satisfied in their work—their postdoctoral appointments—which, by their explanation, appear to share the crucial element of unbridled freedom for research that characterizes the present, which they construe as liberated from the onus of tenure and the performance required for its attainment and also liberated by the idea that now as "free" researchers there will never again be any hurdle as high as the one recently surmounted.

I'd say my postdoc and now are the two highest peaks of my career. Whether this one is higher than the postdoc, I'm not sure. I'm also different than I was as a postdoc, so it's hard to compare. . . . [As a postdoc,] you have no responsibilities; you don't have to raise money; you don't have to teach. Not that teaching is bad, but it's another thing you have to do, and you're completely free to work on your own research. [Now] I'm having a great time . . . I'm working with a great group of people. (9L)

When asked if they would seek an academic career if starting over again, despite the ardor of the tenure rite that above accounts proclaim, they again universally declare in the affirmative, without doubts or reservations. Nor would these scientists retire, as their responses to what is sometimes referred to as the "lottery question" indicate.

> INTERVIEWER: If you could retire now and have the same quality of life as you have, would you?
> SCIENTIST: No, I think physicists never retire. If you make them retire, they keep going back to the department. We can't do anything else except physics. So, absolutely not. (4L)

> No. I like the challenges at the university. I actually don't think I'll ever retire. I might move to sunnier parts of the country, but no, this is who I am. It defines me. (23L)

Or, in a twist on the response, selected scientists would indeed retire if that were interpreted as an opportunity to complete more work, that is, to free oneself of the constraints that some scientists identify.

> I might, yes. Because I could concentrate on my research. I wouldn't stop working, but I'd stop having my students. I would continue my research. A lot of our retired faculty work harder than they ever did, but they're working on their research. (9L)

Moving into mid-career, their passion for science remains strong and steady.

The success these scientists have found appears to operate as an important factor in yielding a strong confidence in the operation of the scientific reward system. The reward system in science distributes recognition to scientists in the form of honors, awards, and esteem. Prior research has extensively examined the reward system of science, with emphasis on whether the system operates according to functional, universalistic criteria as opposed to functionally irrelevant, particularistic criteria (for example, see Clemente 1973; Cole 1992; Cole and Cole 1973; Cole, Rubin, and Cole 1978; Crane 1965; Gaston 1978; Hargens and Hagstrom 1967; Long 1978; Long, Allison, and McGinnis 1979; Reskin 1976; 1977; 1978a; Zuckerman and Merton 1971; for reviews on the subject, see Long and Fox 1995; Zuckerman 1988). This body of research emanates from Robert Merton's articulation of the norms of science, specifically the norm of universalism (Merton 1973c). The norm holds that scientists should judge scientific contributions according to "preestablished impersonal criteria," free from consideration of social attributes of contributors such as their age, race, ethnicity, class background, gender, doctoral or employing institution, or past achievements (Merton 1973c, 270–271). The norm also holds that scientists, like all academics socialized to the role of the production of knowledge, should be rewarded by able judges of the scientific community in ways commensurate with their contributions (Merton 1973a).[2] Here, one is concerned with scientists' *perceptions* of the reward system of science, apart from whether the system is in some sense objectively fair or not. Scientists' perceptions of the reward system and its operation are important because they influence scientists' behavior and enactment of professional roles, including their motivation for continued research productivity.

In the present instance, elites are beneficiaries of reward systems, if perhaps not always in the levels and ways desired at various instances. In sum, they are successful *because* of such systems and their operation. This subset of scientists is thus especially apt to say that systems of reward and recognition in science are fair, even while simultaneously able to see occasions or ways in which they are not.

> I feel sufficiently recognized for my work. [The system of recognition] is basically fair. But I would say less fair compared to promotions in the university. What is recognized as "good work"? This is really, really tricky, you know. We know a lot of famous people before doing very good research, but maybe it was just for some extended reasons. There are maybe other people who made an important contribution and didn't get recognized. That may happen. (5L)

> INTERVIEWER: Do you feel sufficiently recognized for your work?
> SCIENTIST: Yes.
> INTERVIEWER: Do you think the reward system—the system of recognition in science—is fair and equitable?

SCIENTIST: Sure. I mean nothing would look particularly broken.

INTERVIEWER: Do you think it's frequently the case that recognition is delayed in going to people?

SCIENTIST: Yes, that's often true, and that's quite normal, especially in research that is novel. Usually, novelty is recognized only years later.

INTERVIEWER: Are there ways in which you have found an academic career to be unrewarding?

SCIENTIST: Not yet.

INTERVIEWER: Do you anticipate that you might find it to be unrewarding?

SCIENTIST: Everyone, sooner or later, grows up and slows down, and then probably it would become unrewarding.

INTERVIEWER: In the past decade, have you seriously wanted to leave this university?

SCIENTIST: No.

INTERVIEWER: Why is that?

SCIENTIST: I feel good. I'm happy.

INTERVIEWER: And you feel that you have everything you need to do the work you want to do?

SCIENTIST: Yes, pretty much. (15L)

If there is a pronounced concern of scientists of the elite world at this time in their careers, it is not recognition. Instead, it revolves around funding for research. *Funding*, the word that scientists universally use as a kind of code for scientific involvement, success, and continued recognition, is nothing short of a lifeline that enables scientists to advance in a career. In many physics careers, funding is necessary perpetually. Thus it comes as little surprise that scientists greatly concerned about funding are funded. They are concerned about the future of their funding, and hence the future not only of their program of work, but also their daily routines and pleasures.

> It is very important to me to be funded. I don't quite know why because that brings also complexities to one's life, but there's some kind of . . . attachment I have to being funded, some sort of measure of worth. On the other hand, I hate writing grant proposals, and I'm in various structures which provide me with a huge amount of funding, relative to my field. I wouldn't be surprised if I'm funded within the top 2 or 3 percent of people in my field nationwide. Over the past few years, I've typically

had twice as much funding as my peers. Nevertheless, I always feel like it's about to dry up, and I always find that that's something that really bothers me. So that's a constant source of anxiety, and I certainly am very, very responsive to agencies that fund me. . . . I find myself hypersensitive about that sort of thing, just because I'd hate to miss out of being funded just by a hair's breadth. . . . It's a funny issue, which is that anybody who's in a job like mine had some reasonable amount of success over the years, and at some level the pressure is increased because you can only fail at this stage. (23L)

The subject of funding does not go away, neither in the life and work of the contemporary scientist, nor in an interview with one. Later in the interview, on a different set of topics, I asked the scientist what change he would make if he were a university president. He answered this way:

If a university president said, "I can offer you a deal. You're not allowed to seek external funds, but I guarantee you a postdoc for students until you retire," I will take the deal. So for me, anybody who will bring me some unrestricted funds is good. . . . It would be really nice, just be really nice, to take the edge off funding pressures and have a guaranteed $50,000 a year or something like that. If I were guaranteed $200,000 a year, I would take that in a heartbeat. (23L)

For most of this elite subset of scientists, the career progresses either as expected or better than expected when viewed against the backdrop of the previous ten years, when they saw their full-fledged professional careers begin. And while stabilization and rededication characterizes all of the cases of this elite subset, some individuals arrive at this point via a different course, one filled with much greater unease than that which characterizes the standard.

At the time of the first interviews, I asked a scientist—like all of the scientists—how he envisioned his future career, where he saw himself headed, and what he wanted to achieve. The discussion was uncharacteristically tense. At one point, the respondent said: "I'm hesitating a bit because I don't particularly feel so happy with this line of questioning" (2F). It was clear at the time, but even clearer in hindsight with the benefit of a substantially contrasting second interview, that he felt then under extraordinary pressure to perform as a scientist. At the time, with considerable stress and strain, he made these observations:

It's been two years [since I arrived at my present institution].... I have done some good science in the last two years, I've learned a lot. I've made some advances in the field that I perhaps would not have been able to make if I wasn't here. So, scientifically, it's been a good two years. Career-wise, it's hard to tell. I'm at the stage that the last two years have been an "investment two years," getting some techniques going, getting some data, but the results haven't jelled. From a career point of view, it's still in this building phase. If the results go really well, the career will blossom. If the results over the next two years from my current effort don't go so well, then my career will stagger a little bit. (2F)

Ten years passed, and on an arid summer day I found him at work again, this time in a different part of the country. He had gone to lengths to arrange for my visit; he would be the only individual I would interview at this particular institution, situated as it was some distance from the other interview sites. He is one of the four scientists (one of three in the elite early- to midcareer cohort) who moved to a different institution since the time of the first interview, in his case from an elite to a pluralist university. His case, like the other cases of institutional mobility, provides an opportunity to investigate how career patterns are similar and different from those of his cohort.

INTERVIEWER: What have been the most significant changes in your career in the past ten years?
SCIENTIST: Clearly that I went through the tenure process at [my former institution] and was unsuccessful at that, and I came here after that. That would be the major marker.... There were indications toward the second half of my clock there that people were saying the case didn't look too strong. There were some questions about it, but I decided to give it a shot, even though the advice was mixed. There were some people who said, "Things might go well in the last couple years, so the case may strengthen." But other people were saying that, given that there were questions, I should start looking for a job earlier on in the process rather than wait for the full tenure clock to go through.
INTERVIEWER: Why didn't it work out?
SCIENTIST: You know, it's hard to say. I think it's a mixture of we got a bit of a null result from the experiment that I was pushing ... we didn't discover the Holy Grail. There are big searches for a particular effect, and in the experiments we ran, we didn't find that. So there was

a lot of nitty-gritty detail that we learned . . . but we didn't make the notable discovery. And in order to get tenured at [that institution], you need to have a notable discovery, and have senior people in your camp. If you have those two things working, then your tenure case can get quite strong. In my case, it didn't happen. I didn't have the notable discovery, and the senior people in my camp were sort of lukewarm, not wildly positive. It was a disappointment. It's one of these things where you try for something and don't make it. I'm very competitive, so that was a disappointment. In the end, it hasn't led to bitterness. There were many [junior] colleagues who I met while [there] who were disappointed and then became very bitter about the whole process. I was disappointed, but I don't think I've become bitter about it. It still crops up, when someone asks where I was before [coming here], and eventually out comes the conversation that I didn't get tenure at —— ——. So that's still part of my life story, and I still am disappointed about it. (2L)

How does the elite world look to someone who looks back upon it under circumstances such as these?

Six months after leaving, I went to a conference, and I found myself really enjoying the conference because I hadn't ever really noticed how much energy I spent comparing myself to others. In those last two years [there], where I was going through the tenure process, I think I was comparing how I was doing: how I was performing, what impact I was making on the science, what questions I was asking, how the talks were going, things like that. And when I was out of that competitive environment, I was just a lot more relaxed . . . and I wasn't worried about what people were thinking as much. (2L)

Below, as above, the scientist is able to see readily cultural distinctions between the pluralist institution in which he now works and that of the elite institution where he worked previously.

No one here is competitive in the sense that people here are very interested in doing the science, and we're all driven. . . . But it's not the same competition as when you're trying to be the best scientist in your field. It was unspoken, but implied [at my previous institution]. . . . I'm home earlier on a regular basis, and I have much more flexibility. . . . I am traveling a lot [professionally], so, even on a day-to-day basis, I am more flexible and able to spend more time with the family. On a per year basis, I think I make more trips than I would have if I was at [my former institution]. (2L)

The most prominent change in this scientist, however, centers on the teaching role. It is a change, indicative of a more general pattern, that sets this scientist apart from the elite subset of scientists in which he began his academic career. As prior evidence conveyed, his elite contemporaries embrace research principally: it is viewed by them as their primary academic role. In the present case, we observe modulation in those claims.

I think the biggest change would be in the teaching. I don't know where I was ten years ago, but I've always wanted to put a big effort into my teaching. And being here, [at this type of university,] has allowed that more. . . . At [my former institution], I think there is a double standard. There is the expectation that you'd be an excellent teacher. . . . But what I found was that if I put effort into teaching, which I did, and I got a reputation as a good teacher, that somehow matched with people's judgment about where my research was going. I got labeled as somebody who cared about the teaching, whereas the stereotype was that you should be so driven by the research that you would have not much passion left over for the teaching. . . . At [my former institution] the high regard is for the completely driven superstar in research. And here, the high regard is that I can balance. (2L)

The meaning of the teaching role is set in sharp relief when viewed against accounts of other elites of this cohort. For most elites of this cohort, the teaching role was more central in the early phases of their careers, when as teachers they were largely untested, their course preparations were as yet incomplete, and teaching and all of the lived experiences it entails—in and outside of the classroom—was, in short, new. By mid-career, the teaching role is viewed by nearly all of these elites in terms subordinate to research.

Teaching doesn't mean that much to me anymore. . . . I could put more effort and energy in it, it's just not a high priority anymore. . . . I feel like I know what it takes and I can do it, but I don't see cost-effective ways to get better at it. . . . I don't see a challenge anymore. . . . I know what it takes to be appreciated. But to take it to the next level would require investment in time for which I see no real reward. . . . There's nothing really to be gained financially, and it would just take time away from research. (23L)

I don't enjoy classroom teaching. It just takes too much time. You have to prepare the lectures, prepare the homework, prepare the exams, grade the exams, give oral exams to graduate students. There are a lot of time-consuming things, it consumes a lot of time. There are a few positions in [this institution] where you don't have to teach. They just do research all the time. (5L)

But, returning to the prior scientist who switched institutions, one finds different evidence. Having relocated to a pluralist from an elite institution, he is structurally and culturally enabled to assume a changed identity as a scientist—one who emphasizes both research and teaching—the *pluralist* trademark. Yet the evidence suggests further that his identification with the teaching role is pronounced.

Even here [at this university where I now teach], teaching is still not the valued metric, it is not the valued activity that I know it should be. I would raise part of the pay structure, create teaching awards, so your pay goes up. I think that universities should change to give a pay structure that is based on teaching excellence. (2L)

This accounting is less pluralist and is instead more communitarian in form and substance. As a faculty member in a pluralist institution who has adopted a communitarian-tending emphasis and outlook on his work, the scientist may be identified as a "communitarian pluralist," inhabiting that sociocultural space on the continuum of academic worlds where the pluralist and communitarian sectors bleed into each other, producing this hybrid form (see figure 1). His account suggests that, at his former institution, he may have begun developing an outlook and orientation like that of a "communitarian elite"—a member of an elite university who, while generally active in research, champions the teaching role, identifies with it strongly, and partakes in it regularly as a central staple in the constellation of academic roles. These latter qualities now appear more mature and more central in an environment that appears to have made possible their maturation and centrality to this scientist's conception of academic work.

The two other elite scientists who relocated did so to pluralist institutions. But in contrast to the scientist above, their patterns of work parallel more closely the dominant patterns observed of the other elite scientists. Their reasons for relocating also differ from the scientist above (see table 9). One relocated because of ambivalence about the outcome of the tenure decision that would have been made were he to have stayed at his former institution, the other because of a professed "better opportunity," first moving to the renowned Institute for Advanced Study in Princeton, New Jersey, which involved a purely research appointment, then to the pluralist university where he now works, which also is a highly regarded center for scientific research and considered one of the premier public research universities in the United States.

The patterns of these two cases suggest ways in which conditions of careers improve in the eyes of those attempting to develop them. In this sense, all three of these cases of institutional mobility convey more agreeable adaptation to an environment. All three scientists more strongly identify with their present rather than their former institutions, the conditions of work they provide, and the kind of science they are able to do. And they are apt to identify the costs and constraints of having worked where they did previously.

But then the cases diverge: one (above) adopts more of a teaching orientation and outlook on the professional career, while the other two (below) proceed with a research orientation and outlook. What may be the sources of this divergence? One is likely the institutions to which these scientists moved. For the sake of argument, let us say these institutions are the University of Nevada, UCLA, and Indiana University. While all pluralist, these universities inhabit different locations on the continuum of academic worlds. One, Nevada, is closer to the communitarian end of the spectrum, while the others are closer to the elite end of the spectrum, with UCLA arguably closer to that end than Indiana. The institutions provide different structural and cultural grounds for careers, and hence, probabilistically, one stands a greater chance of altering a career orientation and outlook if one were to move from an elite institution to Nevada as opposed to UCLA or Indiana.

Another possible source of the divergence is the reason the individuals moved from one institution to another. In the one case above, however successful the scientist is, and however well adapted he finds himself in his new academic environment, there has been demonstrable failure, at least from a strictly institutional point of view. The two cases below relocate on their own volition and in the absence of this formal sanctioning. There is little systematic evidence on the effects of the denial of tenure on individual academic careers, but one can surmise that this event has the potential to cause significant disruption and even disavowal of previously held beliefs and values about one's work and roles performed. An outcome can likely involve substantially changed outlooks and orientations to work, including the possible reconfiguration of roles and the priorities one assigns to them.

For the two other scientists, the research career blossoms with their moves. They see their research progress and, at least in one of these cases, yield results to international acclaim. One again observes the themes of stabilization and rededication.

> The progress has been steady. Every two, three years, we [the research team] move up a little bit. The change is that now the community pays much more attention, and we needed that. So in some sense we are more relaxed now. . . . The stress has disappeared after the work was accepted in the community. It is now appearing in all kinds of newspapers. It is very much in the public eye. . . . It's not that we have become less productive. . . . Productivity has not declined. We are less stress-driven. . . . I used to work with almost feverish worry [when you interviewed me ten years ago]. I would be working around the clock, I would not like to see movies or go and visit people or take time off for a vacation, ever. . . . I hadn't achieved something [by that point]. Now we are not stressed. So, if it's the weekend, I could take one day off or two days of the weekend and go see a movie with my kids. I would never do that before. . . . Right now, we would have to call ourselves [referring to his group of researchers] absolutely near the top. . . . I feel very satisfied. . . . I would say we are in the middle of our dream. A few years ago, I would have said I dream of solving the black-hole paradox. Now I can say we have solved it. (4L)

> There are no dramatic changes that I see in myself. . . . My work continues to be the same. . . . I used to work on a very different kind of problem . . . but now I work on a different kind of problem, much broader. . . . [I believe I define success] the same way I did ten years ago. . . . Professional success is a success where I feel I have solved an interesting puzzle, a really challenging, interesting [scientific] puzzle. That's the way I thought ten years ago. That's the way I think now. (11L)

A review of the transcript from the interview ten years before bears out the scientist's contention, thus illustrating the preponderance of continuity in his career, now more stable and rededicated morally to the scientific ideal of solving problems established by the scientific community as important and worth solving.

> I want to work on problems which are important problems in the area. ... [Success] is very clear in my mind, it is having accomplished, having understood nature a little better, having appreciation for nature and having understood it a little better. ... Having made important scientific contributions, a person of responsibility, a person who took his responsibility seriously and did the best that he could do. (11F)

PLURALISTS

Leaving the elite world of American higher education and entering another — the pluralist world — one encounters a world whose structural and cultural characteristics establish variant conditions for careers. It is possible to witness *systematically* distinct qualities that situate careers in this type of environment, and systematically distinct outlooks and orientations that characterize the individuals who have attempted to develop careers. By this light, one may see one profession, but many paths in it, speaking of the systematic ways in which a profession is internally differentiated, with felt costs and rewards to both individuals and to the institution of academe.

At middle phases in the pluralist world, careers are passing over rocky terrain not anticipated by the pluralist members of this cohort in the early phases of their careers. At that time, the recognition they sought for scientific accomplishment was, like their elite counterparts, still great, and their orientation to work remained moral and centered on research. Ten years later, careers in most of the cases that make up this pluralist subset have assumed dramatic changes.

Whereas stabilization and rededication characterized the prevailing mode among elites traversing early to middle career, *reversal* characterizes pluralists in this status passage. Like elites, the careers of pluralists stabilize, and chiefly this occurs institutionally, through the mechanisms of tenure and promotion, both to the rank of associate professor and full professor. But, whereas for elites this process was enjoined by a rededication to the institutional goals of science and academe, for pluralists it is characteristically accompanied by a retreat from these goals, reversing an early-career orientation. Ultimately, reversals *destabilize* careers for individuals. Careers may be stable in the institutional sense that they are secure, but prove to be unstable in the eyes and experiences of individual pluralists at mid-career.

The reversal is manifest in attitudes that in some individual cases occur in isolation and in still other individual cases where they occur in combination with one another. The attitudes include: disavowal of research and the reward system of science; greater identification with the teaching and mentoring roles; emphasis on family and non-work life; disillusionment with universities, funding agencies, the administration of science, and bureaucratic policies and practices of academic governance; and partial or complete rejection of the academic career.

One observes these attitudes in the following case, a scientist who in mid-career increasingly questions his role as a scientist, the quality and fairness of the scientific reward system, and, ultimately, his future in an academic career.

> I would say in some ways my research career, ten years ago, was at a peak. I was working with two or three graduate students continuously and two or three postdocs continuously. . . . My attitudes about the job, about me, and about the university have undergone tremendous changes in the last ten years. I've gone from having a fairly large amount of [grant] money, especially for the stage of my career, to having my name on a grant, but not taking any money out of it at all. I'm not sure I want to even submit things to published journals anymore. . . . I'm disgusted by the whole thing. . . . I got tired of getting referee reports [on manuscripts submitted to journals for peer review] that spend a page talking about the bibliography; they were entirely concerned with whether I cited their work or their friends' work, and they hadn't read the paper. I got to the point where at [national] meetings I was telling people, "Please don't reference my paper, if you don't read it, don't reference it." It's a game to so many people, and there are many fools. I didn't do this [go into an academic career] to deal with fools. They don't understand basic things. . . . I went from not having tenure to slowly being delighted with tenure because I can do the right thing. I just came from a class [that I am now taking] that's being taught in philosophy, and I'm rereading my Plato. I can do what Socrates did, I can be a curmudgeon, I can be polite about it, but I can [also] go in and actually try to be honest. I'm in a setting where the last thing people want is honesty. . . . You guys play your game, it's fine. There are more important things in life than getting grants from the National Science Foundation, getting Nobel Prizes even or any of that stuff. That's all just a game. I'm interested in solving problems. . . . I [have] enjoyed my

relationship with my students. . . . I am at a crossroad. I'm not sure, to give you an idea of the kinds of options I'm thinking about: I'd love to be a tutor at St. John's, [the liberal arts college]. I'd love to go to Santa Fe . . . and do real education, not lecturing. I'm enjoying my [philosophy] class; it's almost small enough that I can get to know students. . . . I enjoy advising students. . . . In the environment here, I'd like to see having more seminars. A seminar doesn't mean somebody stands up in front of a group and talks. A seminar means you have a small enough group sitting around a table where everyone can actually participate. . . . I know I could teach high school for a while.

INTERVIEWER: Do you see yourself getting back to research?

SCIENTIST: If you mean publishing papers and going to conferences and advising graduate students, no I don't. . . . What do I care for refereed publications? Like I say, there's no value added as far as I'm concerned for the referee process. Occasionally, it's a good referee report, and I learn something, but it's rare. Usually it's more about power, recognition. And even that wouldn't bother me if there were some legitimacy to it. But often I don't feel there's legitimacy. If I get something from someone who I don't think understands things, insisting that I reference something that I think is wrong, just because it's refereed doesn't mean it's right. . . . I'm not angry about it anymore, I just don't care about it. . . .

INTERVIEWER: How would you complete the sentence, "I am more X and I am less Y compared to a decade ago"?

SCIENTIST: I would like to say I am wiser and I am less naïve. But it could be just the opposite as far as I know. I really am in a very transitional stage. I'm questioning whether I want to be in physics. I've gone a little bit even beyond that. I'm thinking I probably will not stay in professional physics. I want to do something very different. (45L)

In the midst of this disavowal of research and the reward system of science, along with disillusionment of university structures and the academic career, the scientist seeks to reconcile his future, much akin to the ways of his pluralist peers. Asked what he would do differently if starting an academic career over again, the scientist asserts, nearly paradoxically but again echoing his pluralist peers, that he would have "stayed more focused on research," recognizing how in the grip of institutional norms, individuals are in more compatible alignment in enacting their roles as scientists. And in confronting the subject of concerns about the career, the scientist's response again indicates the power of institutional norms and the force of one's socialization to them, perhaps especially observable in the midst of deviation:

> I worry that I'll become one of the deadwood, the people who put in what I consider to be a minimal amount of work to collect a paycheck they don't deserve. That I would not respect. That's a concern. I need to figure out a way where, if I don't want to do the research and I don't want to play the research game, I still have a way of doing something that I regard as very valuable to the university. That may be my biggest—one of my biggest—concerns personally. (45L)

Another of the pluralists raised many of these same themes, giving more stress to some and less to others, but highlighting the attitudes that characterize the career reversal in these middle career phases.

> Ten years ago, which was before tenure, I was pretty focused on getting my lab going and being productive and getting publications out for tenure. I was able to do that. I would say that afterwards, however, after tenure, I definitely went through a period of "now what?" I'd been so focused on getting tenure, there was no chance to look beyond that. Then you get tenure, and it's "what's next?" Do I really want to do this for the next thirty years? Although on the surface things seemed to be going well, I don't think I was really that engaged in the work. I would say in broad terms, I lost focus in research and let that drift. . . . A recent NSF [National Science Foundation] proposal was not renewed . . . Lately, we have written some papers that have come back from the reviewers with negative reviews that are just totally off the mark. . . . I feel I'm fighting an uphill battle because people, if they had a reputation, if they're known by all the colleagues and all the referees, people take their word for it much more. . . . I haven't really pushed to try to move forward in terms of building up a big lab or big reputation. . . . Looking back on it, the greatest trough came after tenure. I wasn't aware of it being a significant thing at the time. It came after tenure. It was like, "Okay, I got tenure, now I can relax." [But] I'm not happy. I don't know why I'm not happy. Work doesn't seem to give me satisfaction. . . . I've had major problems getting myself motivated to just sit down and do [the work]. . . . I look back, and I'm amazed that I've gotten as far as I have, given that tendency. . . . I really was not the kind of person who

wanted to find out about the world and what's going on in it. I had this vision of I wanted to do science, and my parents had always told me, if you're good at something, you'll make a living at it no matter what. It was just tunnel vision. And now here I am trying to be successful and working myself ragged to get a 3 percent raise and realizing that I don't really want to do what's necessary, as some of my colleagues have done, to become successful. It's just not me. . . . I do get a lot of satisfaction from teaching. That's one of the more enjoyable things to me. But that's such a small fraction of what I do. Much of the other stuff is not very rewarding. . . .

INTERVIEWER: What would you say have developed to be the most prominent joys at this point?

SCIENTIST: In the career?

INTERVIEWER: In your job.

SCIENTIST: Teaching, and the feeling when we do make an advance and get it published and get it recognized. But by and large, there's not a lot. There's not a whole lot of enjoyment. I don't do it any more because I love to do it. I do it because it's my job and I want to do well in my job, and I need to do well in my job.

INTERVIEWER: If you were starting all over again, knowing what you know now, would you go into an academic career?

SCIENTIST: No.

INTERVIEWER: Why not?

SCIENTIST: I don't have the right personality for being the pusher and the leader in all these different things. I think I would have made a much more conscious decision to go into something which was more of a group environment, and something which I would have looked much more at the economics of, where it would get me in the end, and where it would allow me to live in the country, and at what age it might allow me to retire. Things like that. I would have paid much more attention to those things.

INTERVIEWER: What have emerged as the most prominent complaints about your job?

SCIENTIST: I dislike the most that I constantly feel I'm behind. There are things that I should have done that I haven't done that are waiting for me to get done. I alluded to this before. It's writing grants, writing papers, spending time with new graduate students . . . all the committee work, writing letters of recommendation for people, researching new areas—possible scientific areas, going to conferences, preparing for conferences, writing referee reports for papers and for proposals. There is just a long list of things. And most of those things, I don't particularly get enjoyment from them.... I'm not alone in saying that it's just not that rewarding to be an academic anymore. (56L)

Like his pluralist peer above, the scientist's career concerns center on his future role as a scientist. Like his elite counterparts discussed previously, the scientist's career concerns center on the future of funding. The scientist thus aptly conveys the more general variety of career concerns—principally about the future of one's professional role and about funding—that are most indicative of pluralists in these middle career phases.

I'm worried that I won't continue to get funding. Currently, I have three graduate students. Two of them are quite good. One of them is awful, somebody I should have kicked out of my lab years ago, and just couldn't bring myself to do it.... I don't feel I'm in any danger of being fired in the near future. But in terms of losing funding, oh yes, I'm very concerned about that. [Especially] in terms of, even more, not having the respect of my peers, I'm very worried about that. And so I'm working very hard to try to keep myself being able to make choices in what I want to do in research, but at the same time, trying to figure out, if that doesn't work out, how I would change, what I would do that would still be able to send my kids to college and at the same time make me happy. Do I go into teaching? Do I try to get involved in administration?
INTERVIEWER: Have you seriously wanted to leave this university in the past ten years or so?
SCIENTIST: Yes. But I don't know where else to go. (56L)

Further evidence of these attitudes in which scientists undergo reversals are found in other pluralists' careers. As another scientist remarked: "I put a lot more value on mentoring, the teaching thing in particular. But also the graduate students. Much more human activity.... Making a difference in other people's lives. It's definitely not the satisfaction of solving some problem and having people say, 'Wow, he is a real brain'" (43L). Another male scientist, now a full professor in his mid-forties, rendered this evaluation:

I'm probably not driven by what's happening at work, but in my personal life. I now have two kids, and I didn't before.... One of the reasons I was comfortable with the academic profession is that I knew

I could put a lot of effort early in my career to become successful, whatever that means, and establish a permanent job. So I had kids. They're great kids. I spend a lot of time with them. I don't put as much pressure on myself academically anymore, not nearly as much. . . . I don't have any grand illusion that I'm going to make a great discovery. . . . If I have a good day with my kids and I have a bad day at work, that's still a good day. If I have a good day at work and I have a bad day with my kids . . . that's not a good day. It was the opposite before . . . it was more important for me to have a good day at work. So the driver, family life, for me is important. (46L)

Indeed it was different before. Ten years earlier, when interviewed as a recently tenured associate professor, the scientist remarked on his career and work in a rather different light.

I'm tremendously competitive in anything that I do. Anything. . . . I'm comfortable standing in front of five hundred people and giving a talk. Most of my colleagues are not. . . . And that makes me stand out, probably for the wrong reasons, but I take advantage of it. . . . When I look back, I can't believe that I was as fortunate and as successful as I was in terms of fulfilling what I wanted to do. Every time I set a goal, I achieved it. I was always driven by goals. I was a very goal-oriented person. When I was an undergraduate, I wanted to get into a good graduate school; when I was a graduate student, I wanted to get a good postdoc; when I was a postdoc, I wanted to become a faculty member; and then once I became a faculty member, I wanted to compete for national awards; and when I won one of those, I wanted early tenure. Now that I have early tenure, I guess I want to become a full professor. I would like to remain an active member of this astrophysics [community]. I want to continue to do good work. I don't want to back off and not take chances anymore because I have security. I would like to be in the National Academy someday. That shows respect of your peers. (46F)

It is evident that these changes are made voluntarily, stressing the agency individuals exert over the course of their work. But agency is always situated within social structures, and in this instance the social structure of the pluralist academic world underscores the capacity to accommodate such individual change. In the presence of this structural flexibility are perceived individual costs when such agency is exercised within the system. For this scientist, the costs are put individually this way:

> Academically—I don't think I have the same standing in the department that I used to.... I don't travel.... I think in all areas of academia, but particularly in the kind of science I'm involved in, you have to become a spokesman for something. You have to be the guy or the woman that everybody calls when they want to know something about your particular niche. In order to do that, you have to be the person that's at all the conferences, and I don't do that. It's just not that important. Now that my kids are older and they have things that can occupy their time... I'll start to travel more, and I won't feel as much guilt associated with it, because I know they're doing things.... If I could go back and do it over again, I would try to find a way to be a more active participant in my research area. I think I probably erred too much, because it was easy to stay home with my kids and send my postdocs and students to give talks. While that made them successful, it didn't help me as much as it should, and therefore I may not be at quite the level that I should be. (46L)

The reversals found at these mid-career points in pluralists' careers are associated with numerous other developmental patterns. Several of these patterns have been indicated by the accounts above. Pluralists profess a desire to have embraced research more aggressively and to stay more focused on it were they to retrace their steps to the present. Their thinking is that, were they to have done so, they would not find themselves in such destabilized conditions of their careers.

The objects of other developmental patterns that arise in these career phases are those discussed previously about elites, but in the case of pluralists they take on dramatically different expression. In middle career phases, pluralists are ambivalent about whether they would again seek an academic career; some say they would, others not. Pluralists are ambivalent about their institutions; some have wanted to leave, others more inclined to stay. Pluralists would, on balance, seriously consider retirement now. Pluralists are uncertain about their aspirations; most find themselves tenuously committed to their roles as scientists. Pluralists claim to be less engaged by work—to work "much less hard" and to be "less ambitious." Pluralists are most apt to identify their *early* careers as most satisfying, this despite the high expectations of performance characteristic of that period but, apparently, more favorably viewed against the contrast of mid-career. Finally, pluralists at mid-career register low professional satisfaction, a stark contrast from their early careers, and a culminating declaration of the conditions in which they find themselves and through which they almost confusedly seek greater meaning.

> Aspirations? I don't know yet. There were a whole bunch [in earlier periods]. I'm really considering a whole variety of things, okay. Aspirations. Well, concretely, I want to resume some of the research that I've been doing. But I would feel that my career wasn't satisfying if that's all I did. I look around the department, and I look at some of the professors who have been here for ten years and longer, and I say, "My god, they're doing the same thing." And I don't want to be doing that. It's also possible, but I'm not that thrilled with the idea, of doing something in physics education research. . . . I'm not dreaming these days. If you asked this question a couple of years ago, I had dreams of unifying the state [of my specialty of physics] in some sort of theoretical institute. But things happened, [and that's gone now]. (43L)

> I'm working less hard than I was ten years ago. I'm probably spending more time just staring off into space every once in a while. It's gone up and down. But I'm probably working less hard, and I probably will always work less hard. I think ten years ago I was more focused without having to think about what I was doing, going through a prescribed set of activities that I was completely confident in. . . . I feel like a lot of work becomes heaped in a meaningless motion. . . .
>
> INTERVIEWER: Has there been a period of your career or an age when you have been the most satisfied?
>
> SCIENTIST: About ten years ago. I really do think about that time. There was stuff going on with the research commission that I was having a lot of fun with—that was a university and national level of doing things that I was really optimistic about. I had three PhD students who were all very good, and I was enjoying the problems we were working on. I was happy.
>
> INTERVIEWER: . . . What would you say are your current aspirations?
>
> SCIENTIST: That's what I'm sorting out. I'll give you one that I re-

ally do think I would be happy with. One of my aspirations would be to retire from here, in six years, [at the age of] fifty-five. (45L)

The scientist below, an associate professor at the time of both interviews, highlights some of the variety found in pluralists' mid-career accounts. While ambivalent about his work, particularly the rate at which he has advanced in it, the scientist nevertheless would seek an academic career again and remains committed to it now and in the future.

I hope [I will be promoted] soon. I've gone through a couple of periods where I've lost [grant] money and I've gotten it back, and I'm getting it back again. . . . I understand why it hasn't happened. It is because of the funding. I really do things that I like to do, and I believe that that is the price I pay. I want to pursue what I want to do. . . . I would have liked to have accomplished more. My experimental program had problems about five years back in terms of graduate students. I would have liked to be further along in the problem I'm interested in now, by maybe two or three years.
INTERVIEWER: . . . Would you say that you're working harder, less hard, or about as hard as you were ten years ago?
SCIENTIST: I think I'm working less hard.
INTERVIEWER: Why do you think that?
SCIENTIST: You get a little tired.
INTERVIEWER: . . . If you were starting all over again, knowing what you know now about academic careers, would you go into one again?
SCIENTIST: Yeah, oh, yeah. Absolutely.
INTERVIEWER: . . . Over the past ten years, have you seriously wanted to leave this university?
SCIENTIST: No.
INTERVIEWER: So it's proven to be an environment in which you are satisfied?
SCIENTIST: Yes. (60L)

In principle, scientists in the pluralist world who depart from the central patterns described above would approximate those of elites or those of communitarians, thus recognizable either as elite pluralists or communitarian pluralists. In this cohort, one finds just one case of such departure, that of an elite pluralist. The scientist establishes this status by two principal ways. First, he more approximately exemplifies the career pattern of stabilization and rededication most characteristic of elites. Second, he relocates to a more elite university that confers on its department of physics a status more elite than at his prior department. The scientist is the fourth to be examined of the four scientists who relocated to a different university since the time of the first interviews. In the case of this scientist, he relocated two years prior to the follow-up interview in order to assume a significant administrative post that enables him to oversee the organization and administration of physics at his university. Like his counterparts who relocated, the scientist interprets his move as a stimulus for further work. Relocation is seen to stimulate the science as well as the individual and career.

> My work habits have stayed really very stable, especially compared to a lot of people I know. In coming here, actually my productivity went up the year I got here, I think because I felt like I had to prove myself over again. I've actually published quite a few papers this year, even with the administrative work I've been doing. I think part of it is I had gotten into this complacent habit at [my former pluralist institution] of just puttering along, doing a little of this, a little of that, and just basically doing enough to get by at the level that I felt was a good, reasonable level, but without really trying to do anything truly outstanding. After I got here, I think because it was a new place, I felt that I really had to prove that I could keep my research career going. I really hunkered down and tried to work at a fairly high level. (59L)

Like elites, the scientist views his career as having progressed as expected or better than expected. His overall level of professional satisfaction is high. He sees himself working harder and perceives himself to be as ambitious, if not more ambitious, compared to ten years ago. His scientific aspirations remain keen. But like his pluralist peers, the orientation to work assumes a more utilitarian, as opposed to a strictly moral, form. Scientists morally oriented to their work engage in scientific research, and other academic roles, as ends in themselves. In general, the utilitarian orientation stresses the ways in which scientists weigh the costs and benefits of their work, seeing it in more instrumental terms. Utilitarianism may be expressed by such behaviors and attitudes as a greater embrace of family and leisure in contrast to work, a deliberate calculation of the economic rewards of work and what might be done to maximize them, a lessening of work commitment and involvement—behaviors and attitudes evident in accounts above and below.

> I think the research part of [my career] has probably progressed [as expected]. But I had no intention ten years ago of getting involved in any kind of administrative work. It just wasn't on my horizon. The administrative [role]—that's been the biggest change and that was not something I set out to do ten years ago.... Overall, I would say I'm working harder [and am] about the same [when it comes to ambition].... I'm certainly not less ambitious. I think my concerns have changed. I'm much more concerned now about making enough money to make the house payments. To do that, I need to be successful in my job, whereas ten years ago, financial issues simply were not even part of the picture. Ten years ago, I was worried about my raise because, and this is common among faculty, the raise reflected the chairman's view of my academic work. When people go in and complain about the raise they've received, nine times out of ten it's not because they need the money. It's because so and so got a higher raise, and they're not doing as much as I'm doing. Ten years ago, I was part of that group—I viewed the raise as a sign of my work as a scientist. Now those kinds of ego issues just don't enter into it. I see it as a pretty raw statement of: do I have enough money to pay the bills or not?... But I'm very satisfied actually.... I'd like to get lucky and make the big discovery still. You throw out these theories, and you hope one of them turns out to be right, and it would be nice to do that. I'm still hopeful.
>
> INTERVIEWER: In learning what you have about academic careers, would you go into one again, if you were starting all over?
>
> SCIENTIST: Oh, absolutely. I think it's the best job in the world. On a bad day, this job is better than most jobs on a good day. I think being an academic scientist is the best job in the world. I can't think of a better job, other than inheriting a lot of money and doing whatever you want.
>
> INTERVIEWER: ... How do you see your aspirations?
>
> SCIENTIST: I would like to continue to do interesting work, and I'd still like to hit the jackpot some day. I haven't given up on that. I want to be successful as an administrator.... I'd like to continue my research. I think I'm less single-minded, that would probably be the best way to put it. I'm less single-minded than I was ten years ago. I still consider

myself ambitious, it's just that my ambitions now are forking in a whole bunch of different directions. (59L)

While pluralists did not foresee in their early careers the extent of the changes that would occur to them at mid-career, it appears they existed then as now in a type of academic environment that would accommodate such change, as to make this evolution a defining, modal characteristic of status passage in the pluralist world. Variety of role emphasis is the celebrated trademark of the pluralist world, with individual practitioners embracing research in some phases of the career, teaching in others, or both simultaneously at still other times. If individuals enter administrative roles, in this world it characteristically comes at the expense of involvement in research and often in teaching as well—career administrators are most apt to be found in pluralist (and also communitarian) worlds of academe (noting that the administrator-scientist above now belongs to the elite). Individuals engage in this variety of role emphasis without strong institutional sanction. Indeed, it would appear to be an acceptable outcome to the mixture of missions also characteristic of pluralist institutions—the mantra triad of teaching, research, and service—which, while celebrated by these institutions, also likely lend themselves to a confusion of goals but, clearly in the end, to plurality.

It is now possible to account for why pluralists at mid-career drop off in their publication productivity in comparison with elites and communitarians in the same phases (table 20). In mode, they undergo reversals, and this experience dampens their scientific output. In the absence of reversal, their publication productivity would clearly climb, and most likely at a rate more approximate to elites of the same age. Stabilizing and rededicating themselves at mid-career primarily to the institutional goals of science, elites accelerate in publication output while pluralists, reversing these orientations and outlooks, decelerate in their publication output. Communitarians, the final group that comprises the representation of academics, academic institutions, and the careers that link them, find themselves at mid-career in patterns different still. It is they to whom I now turn.

COMMUNITARIANS

Perhaps most striking about the communitarian world of academe, evinced when I interviewed scientists in this sector originally and again for the follow-up study, is the manner in which individuals regard their employment. For almost all communitarians, their institutions are understood and described to others as places at which they never expected to end up. The customary rhetoric is as if higher education institutions may be classified dichotomously in a simplified scheme: institutions that individuals strive to enter and institutions that individuals strive to leave. Much of the communitarian career appears to be spent explaining, at times almost apologetically, how one arrived at one's present institution, the problems posed to careers that one has encountered in this world, why individuals have remained, and with what costs and benefits. One does not hear this kind of accounting from elites nor from most pluralists. Most communitarians, particularly members of the youngest cohort, did not anticipate employment in this sector of academe, and the fuller accounts of their careers in this world profess that people stay because it has come to entail secure employment and because they have nowhere else to go: they have been promoted into immobility.

> The reason I'm staying is because I'm unhappy, but there's no place to go. I mean everybody in our department feels the same way. They wouldn't be here if they could find a job elsewhere.
> INTERVIEWER: If you could go somewhere else, would you?
> SCIENTIST: Of course. Yes. But that's such a remote possibility. I don't even think about it [with any practical seriousness]. (30L)

This reality is most keen at early career, perhaps because the contrast between the past and present is at its greatest, that contrast between one's graduate and postdoctoral institutions and one's present institution, between the grand aspirations inculcated and developed in graduate and postdoctoral education and the more muted aspirations leveled by the realities of work in the communitarian world.

> I thought I would end up in a better, main university, like Yale or Harvard or someplace like that. That was as a graduate student. I would have thought, well, if you go to [this university], my god! Are you idiotic? People there won't be doing anything. It would just be a complete reversal of everything. I think when you apply to these [elite] places, as I have, and you don't get an interview, you start to wonder, gee, did I really do something wrong? But then when you go to meetings and conferences where the topic of looking for a job comes up, you realize that even places like the University of Southwestern Texas, when they

send out an advertisement for an assistant professor, they get about four hundred applications . . . and they accept only one person. . . . It's simply that there are too many . . . PhDs for the number of academic positions. (30F)

You always make choices because you can't do everything. I probably could have looked for an industrial position. There are other positions I could have applied for along the way . . . I don't see a future. I don't see the university living up to its potential that I saw here when I came.

INTERVIEWER: . . . Is this the type of university where you thought you would end up?

SCIENTIST: No, not really. . . . I thought, [for example,] Kansas State, there are good state-funded schools around. Schools like that is the kind of arena I thought I would want to look at when I took an academic position.

INTERVIEWER: Do you regret not having gone to one of those types of schools?

SCIENTIST: Those opportunities never really existed [because of the saturated academic job market]. . . . The question here is, what are we going to do [as an institution]? If we have to hit the ball and drag "Jim," I don't see much of a future here. When other people are playing a foursome [collaborating, and connected to political cliques that control institutional resources], they can get around a lot faster than if you're dragging this dead body around. What I don't know, and what I don't know how to find out, is what the future is going to be. (32F)

What *has* the future entailed for communitarians since first interviewed in their early careers? Whereas for elites at mid-career the modal career pattern consists in stabilization and rededication and for pluralists it consists in reversal, for communitarians it consists in what one may most aptly call *stasis*: a continuation of the same leveling off of the career, begun in early phases, that produces a prevailing sense of stagnancy among communitarians at mid-career.

Like elites and pluralists, communitarians at mid-career have attained security in their positions through academic tenure and promotions. Yet a divide between professional aspirations and opportunity to realize them within their institutional environments—a phenomenon recognized early in their careers—continues to deepen through mid-career and to blunt expectations about the future. The resulting outlooks and orientations to work typically consist of a blend of frustration, cynicism, and resignation. In their roles, communitarians embrace the academic triad of teaching, service, and research, but they give more emphasis to service and to teaching than either elites or pluralists. At least rhetorically, the triad is broken up by communitarians into roughly equal shares, which serves as a symbol of this world. Their records and rhetoric show that they continue to do a little bit of everything under the rubric of the academic role: they continue to teach, to serve in their departments and institutions on committees, and to engage in what typically is a modest program of research. These routines, which appear to lose excitement, enjoined by the leveling off of aspiration, produces the stasis characteristic of this phase in the communitarian world.

> [My outlook] still gets me frustrated in the environment. . . . This institution, more than any other I've been at, rewards politics rather than performance, and I've always been a performance-type person. It's frustrating. I end up spending a lot of time dealing with issues that I probably shouldn't have to spend any time with at all, such as dealing with unfair treatment, and it's not just me, it's a lot of people. . . . I probably could have had a more active research program. I certainly got bogged down by the department. I just don't know how I could have bypassed some of this political stuff, because it's so tied into your life. . . .
>
> INTERVIEWER: Would you say that your career has progressed as you had expected it would?
>
> SCIENTIST: No, I wouldn't say that. I certainly have had a lot of distractions around here, and I think I could have been much more successful. . . . I think there's a lack of support, actually obstacles. I think there's been an orchestration of people not wanting people to succeed, not wanting to succeed in the department because there are things they can't do. I see it happen to other people. . . . I'm less focused on research partially because of what I just said. . . . I'm probably less ambitious than I was ten years ago, maybe more pragmatic. . . . I'm doing more service-related work than I was. I'm on a lot of university-wide committees. I'm on the executive committee and chair of the senate budget committee. . . . I've been frustrated in some ways. If people put up enough impediments, that can affect what you can do. I think I've always been capable of doing much more. . . . There's a Gary Larson cartoon that seems to fit. It's a caveman sitting there, and he sees this

bird flying over. He starts flapping his wings, and just can't get off the ground. In the next frame, he's got a bow and this dead bird with an arrow through it sitting on the ground. I see that around here. (32L)

When I started out, I was setting up to do research and was teaching two or three classes a year or something like that. . . . I've been sidetracked in the intervening years because I was asked to serve in a couple of different ways in very large service roles. And so, as a consequence, I would say that my research has slowed down as far as I can envision. . . . I would suggest that maybe I would be satisfied perhaps with a little less ambitious measure now in terms of research accomplishment. . . . I still would like to do some things. I guess it's a little hard to say. I'm probably less ambitious in some regards. . . . I haven't been terribly productive in the last couple or three years from a research standpoint. That's probably the biggest single thing I regret today. I just haven't spent the time at it. I guess, if it were really a priority of mine, I would have forced myself to spend more time at it. . . . It's always been a goal of this university, since I've been here, to become sort of a preeminent research institution in the region. But that was always a goal. Will we ever get there? I don't know. People have been saying that we're trying to move in that direction, but I don't know if we've moved a whole lot. . . . I really do like the teaching. I love the interaction with students. I do like that, there's no denying it. I like interaction in the classroom. I like interaction in the laboratory. [But I find that] you're happy to interact with so many students that it begins to dominate everything you do. . . . Still, I really like the student interaction. (29L)

The accounts above illustrate the more general patterns of deceleration in research, greater identification with, or at least involvement in, teaching and service roles, and, particularly in the first case, continued frustration, resignation, and cynicism about the changes that the institutional environment has brought about in individual careers.

The development of stasis as the modal career pattern is accompanied by several other behavioral and attitudinal characteristics of communitarians in these phases. Many of these characteristics are evident in the accounts above. These are the dimensions along which the careers of elites and pluralists have been analyzed; among communitarians, the characteristics assume still variant expression.

Specifically, communitarians at mid-career tend to believe they work less hard on the whole and are less ambitious. They claim their overall satisfaction is low or moderate at best. Yet, as if in a twist, many identify their peak career satisfaction presently, a time which, as it turns out, their aspirations and expectations for the future are, more than ever before, in greater alignment with local opportunities to get ahead, thus reducing comparatively the degree of frustration and anxiety. Alternatively, other communitarians identify their peak satisfaction as occurring prior to the onset of their present jobs, usually their postdoctoral appointments. In both cases, however, the responses are subdued and unresonant, as if a question about "peak satisfaction" was not particularly salient for these scientists.

Further, communitarians have wanted seriously to leave their present institutions. Most would not again seek an academic career or, if they did, would try to enter what in their words would be a "better" institution—one in which there is greater professional opportunity, especially in research. Their professional aspirations for the future are substantially dampened, and in some cases nearly eradicated, an extension of a pattern begun within their first years as professors in this world. Finally, in a pattern wholly unique to communitarians that was first observed even in these scientists' early careers, they voice serious desire to leave academia for alternative employment, some of which bears no resemblance to their roles as scientists or university professors.

Overall, these patterns were first discovered in the foundational study, when these scientists were assistant professors or early associate professors. The patterns have continued and deepened over the intervening ten years. Thus, while the scientists may be at markedly different phases of their careers, they continue to experience and be defined by many of the same phenomena that marked, and often plagued, their early careers—an observation unique to the communitarian subset of this cohort. This constellation of patterns is evident in illustrative accounts below, the different accounts emphasizing the elements of the constellation variously:

> It's quite clear that life is finite. I realize that now more than before, and I think I'm a smart person, everybody tells me that. I'll probably go into business . . . I figured out that that's probably what I would do, go to business school. I'm actually thinking of going to business school, night school at ——.
>
> INTERVIEWER: . . . In learning what you have about academic careers, would you go into one again, if you were starting all over?
>
> SCIENTIST: No.

INTERVIEWER: Why?

SCIENTIST: Because the money is not right, and it's a lot of work. Initially, it was fun because, after all, if you're a grad student, that's what you want to do, you want to be in academics. But then you have your fill of academics, and you [realize] the pay isn't that great . . . it's also a very crazy type of existence. I have to work until nine or ten in the evening sometimes, and I have to come in at odd times, whenever an experiment is running. I'm teaching a graduate course now that is taking up all my time, and I just find that normal people have different lives than people in academia. . . . Also, it's not rewarding in the sense that you don't really contribute that directly to society. You contribute indirectly through research that may have an application in thirty years, or something like that. But it's not like you're building a better anything that has immediate consequences. So these things: you can't leave your work behind, and the results impact society indirectly. It's not like you're selling food or you're building a road, or you're doing something with immediate benefit.

INTERVIEWER: What do you think you would do?

SCIENTIST: As I said, go into business. (30L)

INTERVIEWER: At what point in your career would you say you have been the most satisfied?

SCIENTIST: I don't know.

INTERVIEWER: If you were to identify a high point.

SCIENTIST: Probably when I was a postdoc, I guess. I felt valued, and I don't always feel valued here. So I felt valued. People that I knew then, I still receive recognition from them.

INTERVIEWER: In learning what you have about academic careers, would you go into one again if you were starting all over?

SCIENTIST: I don't know. I'd probably look at different things. . . . Whatever I did, I would look more at the people and what the goal really is. . . . I'm not sure. I probably wouldn't be so afraid of going into an industrial job now with the background I have.

INTERVIEWER: . . . What would you say have developed as the three biggest complaints about your job?

SCIENTIST: Dishonest colleagues would be number one. An administration that supports them by supporting politics over performance, and the discouragement that's put in place by those two aspects of it.

INTERVIEWER: Would you say you feel frustrated now?
SCIENTIST: I'm frustrated.
INTERVIEWER: Do you see this changing?
SCIENTIST: If I didn't, I'd shoot myself, or somebody. I hope it changes. I'm going to have to fix it or let go. (32L)

A scientist at another of the communitarian institutions, the one that lacks a doctoral program in physics, echoed a similar mixture of themes.

INTERVIEWER: In looking back over this time period, do you think your career has progressed as you had expected?
SCIENTIST: I have thought about this often. I'm satisfied. I'm pretty happy. But it could be better. . . . I should be happy with myself being at the university. We've got some support, but not the support that we really should get. . . . We have limited internal resources.
INTERVIEWER: Does that have a bearing on your work?
SCIENTIST: I think so. It limits. It has limited our ability for sure. (33L)

But, again illustrating the stasis and immobility that comes to characterize the communitarian career, the scientist added:

When I was younger, I thought going to a bigger, better, well-known university would help my career. But at this point, I don't think it would really make a difference. For one thing, I'm older. . . . I don't want to go through this whole thing again. You want it to be easy, to take it easy in that regard. (33L)

A colleague in the same department continued the line of thought:

INTERVIEWER: . . . If you were starting all over again, what would you do differently?
SCIENTIST: I would try to find a place that has a doctoral program. That's important. That would provide me flexibility to do certain things. That would make a difference. [Students] stay with you for a longer period of time. You can work on longer things.
INTERVIEWER: . . . What would you say now are your current aspirations?
SCIENTIST: I don't know. This is hard. I don't want to attain

anything. I just want to make sure that I am able to continue to impart good physics training on students, both at the undergraduate and graduate levels.... I would not mind investing more time into undergraduate aspects of physics.

INTERVIEWER: ... Would you say you have a dream in relation to your work?

SCIENTIST: To recruit as many physics majors as I can. That is my big desire. It's not easy, because it's a hard subject. I would like to continue some research, because you must be current. You may not have the highest journal article that you wanted. They reject your work, throw it back at you. That's fine, very fine. But you have to still be current. If I meet you another time, [the question I will ask myself is:] "Am I current or not?" It doesn't matter if I have a research article published; I might be current in the literature—that is important. (36L)

In addition, communitarians at mid-career adopt a specific behavioral and attitudinal response toward the scientific reward system, to which the scientist above begins to allude. At mid-career, communitarian scientists generally deem the system unfair and underscore the ways in which they are disadvantaged in reward and recognition by virtue of their institutional location. The two scientists quoted below illustrate this pattern.

You have to work harder to convince whomever ... needs convincing. Articles for publication, for example: it's not easy getting yourself accepted if you are writing from a place which is not Chicago or New York. You're writing from a place like this one. We have to prove ourselves more. You're often not successful [at a place such as this] in being able to write articles in super-premium journals. We can write articles in medium-premium journals. Publication in the premium journals, I think, is a lot related to the name of the institution. . . . I'm not suggesting every paper submitted by people at top institutions is acceptable at higher-end journals all the time, either. They [the journals] are selective. But it's a clubbish mentality. What we do [increasingly] is submit articles to European journals. And we are met with surprisingly better success. European journals are a lot more open. None of them seem to have this club-like mentality. The reviews are more collegial in nature. They will look more at the content and give you feedback. (36L)

I haven't gotten funding, big funding. Maybe I can get it. And to do that, I need to write a really shocking paper. I think I did a pretty decent work, competitive with a physicist, a condensed-matter physicist at Harvard. I don't feel much inferiority to the professor at Harvard. [But] I'm always frustrated why, you know, he did something, I did something. It's almost comparable. Why he is a Harvard professor, and I'm not? I simply look at what we did. He did, [and] I did . . . creative, original papers, several of them. I don't see it. The quality of my paper is at least as good, not less, than his paper. But I know human society. Politics is a very important part. (41L)

In mode, communitarians at mid-career continue to grapple with many of the institutional conditions first confronted in their early careers. Instead of disappearing, lessening in intensity, or being replaced by different concerns, the conditions—and the sentiments they prompt about work, career, and institution—have deepened into a mid-career maturity. The accounts portray the resulting professional dissatisfaction and discontent with the institution of academe and also the institution of science.

Like the elite and pluralist worlds, however, the communitarian world contains a measure of internal variety—cases that deviate from modal patterns as to resemble patterns more typical of adjacent social worlds. Like the elite and pluralist worlds, the extent of internal variation is limited within the communitarian world, in part a function of the subsample size, but also representative of the proportion of variation one would likely find were the subsample larger. In this instance, there is one member of the communitarian world who differs significantly from modal patterns. The scientist deviates, in both objective scientific record and subjective accounting of the career, as to more approximately resemble elite patterns. The scientist assumes the status of an elite-communitarian.

This scientist, a female full professor now in her late forties, held elite-like status in her communitarian world at the time of the first interview. She was then, and has remained, highly productive in research. She is the most prolific publisher within the communitarian subset of the early- to mid-career cohort, raising the maximum number of publications for communitarians (94) above that of her pluralist counterparts (68), along with the respective average number of publications for each cohort subset (50 for communitarians compared with 44 for pluralists—see table 20). While her publication productivity stands out among communitarians, which in turn makes them stand out compared with the pluralist subset of scientists in the same career phases, her case does not alone explain the productivity inversion between communitarians and pluralists of this age-set—this is more fully accounted for by the reversals pluralists undergo, explained in the previous section, which effectively dampen their scientific output.

At the time of the first interview, her orientation to science and outlook on her career were unambiguously clear and unambiguously elite-like:

> I certainly was always shooting for the highest position. I just wanted to keep going up and up and up. . . . I'm an associate professor now; I've just been promoted this year, and I don't want to stay there. I just keep going up as high as I can. . . . [My aspirations] keep increasing. . . . I keep wanting more. I keep wanting to get further and further and further.
> INTERVIEWER: . . . Do you have a long-term dream?
> SCIENTIST: Yes, I want to win prizes. I have my eye on one of them already. . . . The ultimate goal is to be more and more involved in research and get more and more positions of responsibility.
> INTERVIEWER: . . . Are you ambitious?
> SCIENTIST: I'd be lying if I said I'm not.
> INTERVIEWER: . . . What is your ambition?
> SCIENTIST: It's to keep climbing all the ladders and go up to the highest one. (42F)

In the intervening ten years, the scientist went on to win numerous awards for research, and has become—as hoped for—more heavily involved in national positions of increasing responsibility. Her record, in particular her publication productivity, continues to stand out among her peers, much as her subjective accounting of her career. Ten years later, while the curriculum vitae conveys substantial change of increasing publication, reward, and recognition, the accounting of career remains highly consistent with that stated a decade previously:

> I have progressed in my career. I have obtained the highest distinction one can get in the university. I've also been elected to the American Physical Society as a fellow. I've had many awards. . . . [My career] has progressed very well, perhaps more than I would have thought. I was able to raise a lot of funds. I just had one grant of $900,000. . . .

I actually get the highest grant as a single P.I. [principal investigator] in the [specific grant agency]. In addition, I was able to get several initiatives—big chunks of grants. And that's a competition across four sciences. In '96, I got a million-dollar grant, and then last year I got $400,000. That was a competition in materials science—very few grants were given. So I think I'm doing actually quite well.

INTERVIEWER: Do you believe you define success differently now than you did before?

SCIENTIST: Yes, we get greedy, unfortunately. . . . Right after I got these grants, I wanted to get more. Bigger grants, lots of papers. *Phys. Rev. Letters* [Physical Review Letters] is the best journal. I now have many papers in *Phys. Rev. Letters*, and that was also an objective I wanted. (42L)

Like elites, she exemplifies many of the associated attitudinal and behavioral patterns associated with rededication.

I'm quite satisfied in my career. In ten years, I've done a lot. I've more than doubled my salary. As I've said, I've gotten the highest distinction you can get here.

INTERVIEWER: At what age would you say you've been the most satisfied in your work?

SCIENTIST: Perhaps it's now. This is a good time. . . . I'm more relaxed about things. Before I tended to be a slave driver and not have understanding or patience with people in my group or people who work around me who weren't as dedicated as I was to scientific adventures. I'm happy to be that way. I'm much better at understanding. I wouldn't want to be one of those very successful people who is also very insensitive to their surroundings and people around them. I'm happy about that, I'm more relaxed.

INTERVIEWER: . . . In learning what you have about academic careers, would you seek one again if you were starting all over?

SCIENTIST: Yes. I think academics is what I like the best.

INTERVIEWER: . . . Do you see yourself having a dream?

SCIENTIST: I feel that I had a dream, and I realized it. It's not easy to be a successful physicist. I wanted to be successful. I wanted to be a successful physicist. I had that goal. I've accomplished that. . . . I still want to continue playing physics. I still have good ideas that I'm working on, that I'm developing. (42L)

The case above depicts not only deviation in career patterns but also career patterns of women scientists. The other woman scientist of the cohort is located in the elite academic world. Hence, recalling that just four women are in the sample of scientists studied (and recalling that this represented an oversampling of women physicists employed at U.S. graduate degree-granting institutions), half are in the youngest cohort. Other research takes as its central concern the subject of gender and scientific careers, adopts larger samples of women scientists, and is thus in a better position than the present work, in light of its different concerns, to assess this relationship (for example, see Cole 1979; Eisenhart and Finkel 1998; Fox 1995; 2005; National Research Council 2001; Preston 2004; Sonnert and Holton 1995; Xie and Shauman 2003; Zuckerman, Cole, and Bruer 1991). While generalizations about gender and science cannot be made from this study, some suggestive insights arise in examining the careers of women scientists over time.

First, all of the women scientists in the sample were highly dedicated to the research role at the time of the foundational study. At that time, two of the women were in early career, one was in mid-career, and the fourth in late career. Ten years later, they remained highly dedicated to the research role, as the passages above illustrate, pointing to a consistency over time in career identification.

Second, and again reinforcing a pattern first observed in the foundational study, the women scientists, regardless of the academic world in which they work, are elite-like in their work attitudes and publication productivity. That is, following patterns among elites, their professional aspirations intensify over time. They customarily seek substantial levels of recognition and maintain a moral orientation to work consistent with a focus on research and an embrace of the institutional goals of science. Moreover, across the cohorts in which the women are arrayed, they are apt to report that, ten years later, they perceive themselves to be working as hard or harder than before, this despite the advancement of their professional ages. Over time, they maintain external definitions of success and a positive attitude toward work. Funding stands as their most prominent career concern. They perceive their career to have progressed as expected or better than expected. They claim they would do little differently and have not entertained a desire to leave their institutions or to retire, except, again paralleling elites, if such a change would afford greater time for their own work. Their overall satisfaction, as captured in the accounts above, is high. In the words of the foundational work: "In performance and in narrative, the range of variation of women is slim, narrower than the variation range among men. If not elites in the institutional sense, all of the women are elite-like. . . . It is the most ambitious men whom the women most closely resemble" (Hermanowicz 1998, 186). These observations remain consistent and are reinforced by the longitudinal data. The foundational data suggested, and the longitudinal data lend further support to the idea, that to succeed in science women scientists are held to higher bars, a pattern more fully documented in studies that specifically examine gender and science (Sonnert and Holton 1995). A major consequence is that women are seen to be highly research-oriented and focused on the role that most directly determines their success, all of which are observed, in this study, to endure over time.

SUMMARY

The careers of scientists passing from early to middle phases prove to be markedly different across the major social worlds of academe. As it will turn out, if there is a point at which scientists across the academic worlds shared elements of their careers, it is in their early careers, especially in their very first years as professors, when their outlooks and orientations were most similarly and morally centered on research, the anticipation of great recognition, and the active pursuit of intense professional aspirations. As their early careers develop, one can begin to detect differences in outlook and orientation, described at the outset of this chapter and conveyed in table 21. Ten years later, at mid-career, differences among the groups of scientists have developed further, to the point at which there are now several stark contrasts in what it means to be a scientist and to have an academic career in an American university.

I codify the findings of how the scientists make passages from early to mid-career by using the principal themes arising from the interviews. It is also possible to begin to build a stock of knowledge about the development of careers across the social worlds of academe by bridging these findings with those generated by the foundational study. Table 22 accomplishes just this. In it, one may observe the findings situating the early- to mid-career passages across the academic worlds. The findings are codified by the principal themes, or career dimensions, generated by the study. I seek to build on the knowledge generated previously about these careers by using several career dimensions that are equally relevant to both the foundational study and the longitudinal study. Thus, in the first part of table 22, findings are presented to bridge patterns uncovered between the foundational and

TABLE 22. Early- to mid-career patterns of scientists

	Early- to Mid-Career Comparisons		
Career Dimensions	Elites	Pluralists	Communitarians
CAREER FOCUS			
In Early Career	Research	Research	Research
At Mid-Career	Research	Teaching/ Mentoring/ Research	Teaching/ Service/ Research
PROFESSIONAL ASPIRATIONS			
In Early Career	Intensify	Rescaled	Diminish
At Mid-Career	Intensify	Diminish	Subside
RECOGNITION SOUGHT			
In Early Career	Great	Great	Great
At Mid-Career	Substantial	Average	Minimal
ORIENTATION TO WORK			
In Early Career	Moral	Moral	Moral
At Mid-Career	Moral	Moral or Utilitarian	Utilitarian
WORK/FAMILY FOCUS			
In Early Career	Work	Work	Work
At Mid-Career	Work	Family & Work	Family & Work
ATTRIBUTION OF PLACE			
In Early Career	"Burden"	"Happy Medium"	"Stymieing"
At Mid-Career	"The Best"	"Den of Confusion"	"A Job"
OVERALL SATISFACTION			
In Early Career	Medium	High	Low
At Mid-Career	High	Low	Low

(continues)

TABLE 22. (continued)

Career Dimensions	Additional Mid-Career Patterns		
	Elites	Pluralists	Communitarians
CAREER PROGRESS	As Expected; Better than Expected	Not as Expected	Not as Expected
WORK INTENSITY	As Hard; Harder	Less Hard	Less Hard
OBJECT OF SATISFACTION	Research	Teaching/ Mentoring	Teaching/ Service
PEAK SATISFACTION	Present; Postdoc	Early Career	Present; Postdoc
REWARD SYSTEM	Fair	Unfair	Unfair
DEFINITION OF SUCCESS	External	Uncertain	Internal
WORK ATTITUDE	Positive Negative	Preponderantly Negative	Preponderantly Negative
PROMINENT CONCERNS	Funding	Future Role; Commitment; Funding	Professional Opportunity
ACADEMIC CAREER AGAIN	Definitely	No; Maybe; Yes	No
DO DIFFERENTLY	Very Little; Nothing	More Research Focus; More Aggressive	"Better" Institution
LEAVE UNIVERSITY	No	Yes; No	Yes
RETIRE NOW	No	Yes; No	Yes; No
OVERALL MODAL PATTERN	Stabilization & Rededication	Reversal	Stasis

longitudinal studies. In the second part of the table, new findings generated by the longitudinal study are presented, which aid in understanding the differences that have emerged by mid-career among the three groups of scientists.

Elites, pluralists, and communitarians diverge in the focus they bring to their careers in middle phases. Whereas in their early careers they focused on research, by mid-career this focus differentiates: pluralists embrace teaching and mentoring, in combination with some research; communitarians place greater stress on teaching and service, in combination with a level of research; elites continue to show a primary focus on research.

The professional aspirations of the scientists began to diverge in their early careers, as pluralists and especially communitarians found themselves in environments that constrained their research ambitions. This differentiation intensifies into middle career phases. While all of the scientists experienced a recalibration of their aspirations and "dreams" for professional success, apparent in remarks about becoming "more realistic" or "more pragmatic," elites continue to mark themselves by intense professional aspirations: the individual hope and the institutional mandate for great achievement remains alive, and thus the recognition they seek at mid-career remains substantial, however more realistic or pragmatic they may now be about expectations for their professional future. Aspirations among pluralists diminish, and they come to strive for average recognition rather than great scientific accomplishment. In communitarians, professional aspirations have begun to subside, even at these relatively young points in mid-career. They have come to seek minimal recognition from the scientific community for scientific achievement, a direct outcome of their assessed likelihood of making highly significant achievements within the constraints of their institutional environments.

To this end, elites maintain an overall moral orientation to their work—science remains an end in and of itself, and the pleasure (and success) they find in their work sustains a commitment, involvement, and productivity. Engaged in research, though in lesser degrees, pluralists become more utilitarian in their work orientations. They have become more apt to weigh the costs and benefits of scientific engagement and are increasingly interested in the material benefits that their occupation can confer. Communitarians, too, progressively view their work in utilitarian terms, but with a decidedly less optimistic outlook. For many of them, science is no longer a calling, passion, or profession, but a job.

The changes above are accompanied by the relative stress given to work and family in the accounts scientists provide about their careers. Elites continue to emphasize work. Pluralists and communitarians increasingly emphasize family as a sphere with which to view work in combination. Thus, where for scientists the salience of aspiration and the quest for professional recognition declines, the salience of family increases.

The ways in which scientists have experienced work lend to differing attributes that they attach to their institutions. In their early careers, elites regarded work in their institutions as a "burden," attempting as they were at this time to achieve success in the form of job security and promotion in the career. Having done so at what are among the world's premier research universities, and thereby validated in their scientific achievements, they now regard their institutions—as well as these phases of their careers—as "the best." There is little sense, in the elites' accounts, that conditions for a successful career in science could be any better. Among pluralists, the place of work—and the phases in which they find themselves—were once viewed as a "happy medium"—an academic world in which one could emphasize a plurality of interests and emphases in the academic role. Now, however, it is viewed more aptly as a "den of confusion." While pluralists are secure in their jobs and have been promoted up the ranks, their world of work no longer appears in their eyes to provide strong indication about an institutionally desired role. Thus pluralists proclaim desires to "get back in the swing of things" and question what it is they will do in the future in their search to find direction. In their early careers, communitarians realized that their institutional environments were "stymieing," Ten years later, they view their place of work and this time in their careers as "a job," having grown still more resigned about the comparative lack of opportunity for substantial scientific achievement.

The divergent realities in which the scientists find themselves lead to distinct overall characterizations of their satisfaction. For elites, it is "high," largely because they are now validated members with secure positions in institutions that provide conditions for continued scientific achievement. For pluralists and communitarians, overall satisfaction is low—for pluralists primarily because of their confusion about their present commitment to and future role in science, and for communitarians primarily because of leveled aspiration.

The career accounts thus assume different forms. Accounts provided by elites stress their "becoming," that is, the ways in which their careers and their professional status are progressively changed by scientific achievement. Accounts provided by pluralists and communitarians stress their "being," that is, the ways in which their careers and their professional status do not progress or change substantially due to a comparative lack of scientific achievement. Pluralists provide indication that the form of their accounts may change in later career phases as they rediscover and renew a commitment to science, whereas communitarians provide indication that institutional conditions may never allow such comparable change to occur.

Several additional mid-career patterns are evident from reinterviewing the scientists. These patterns vary with respect to dimensions of academic careers that arose as prominent themes in the interviews. Elites find that their careers have progressed either as expected or better than expected, whereas pluralists and communitarians find their careers not to have progressed as expected. In their work intensity, elites see themselves working as hard or harder compared to ten years ago, while pluralists and communitarians view themselves as working less overall.

At mid-career, the object of overall satisfaction is research for elites, but increasingly teaching and mentoring for pluralists and teaching and service activities for communitarians. Viewing their careers, elites claim that their peak satisfaction is at the present or during their postdoctoral appointments, which in their minds contained a freedom that they have once again experienced at mid-career. Peak satisfaction for pluralists is typically said to be in their early careers, prior to the confusion about their roles characteristic of the present middle phases. For communitarians, peak satisfaction is seen either at their postdoctoral appointments—when their aspirations and expectations for the future were perhaps most keen—or at the present, which, while posing challenges of adaptation and accommodation, appears more resolved than their early careers.

Elites at mid-career are the most apt to believe that the scientific reward system is fair, in part no doubt because they have been significant beneficiaries of it. Even though at mid-career we see elites grow more independently minded, their work and motivation remains attuned to external audiences and judges. In mode, pluralists and communitarians find the scientific reward system unfair and are quick to identify its imperfections. This is accompanied by a changing definition of success, wherein pluralists grow more uncertain about what "success" means, particularly in light of their changed scientific commitment and engagement, while communitarians adopt a more internal definition: they increasingly scout out the means, afforded by their institutions or simply by themselves, to render themselves meaningful in some way.

The overall work attitude among elites is positive: they continue to embrace their professional roles, especially the scientific research role. They now embrace their institutions, and their institutions appear to embrace them. Attitudes toward work among pluralists and communitarians are preponderantly negative. For pluralists, work proceeds under a confusion about the present and future mix of roles. For communitarians, work transpires amid distracting institutional and departmental work conditions, particularly in the form of a politics that communitarians see as endemic to this academic world.

The varying conditions across the academic worlds, and the divergent experiences of work found in them, prompt systematically distinct career concerns that individuals find most salient. For elites, it is funding, since funding is attributed by them as the most critical link to their work. For pluralists, the most salient career concerns revolve around their future role in science and in academia generally—what role this will be, how they will enact it, and with what success and consequence to themselves. They also voice concern about their present commitment to scientific research, inextricably tied to uncertainty and ambivalence about their future roles as scientists. As if to compound the concerns further, they realize the significance of funding and how, if removed or reduced, the other concerns grow overwhelmingly or are eliminated altogether, depending on one's point of view. For communitarians, the most salient concern evident about their careers revolves, almost unceasingly, around professional opportunity.

Would these scientists seek an academic career again were they able to do so with the knowledge they now possess about the profession? The responses differ systematically. Elites definitely would. Pluralists, owing to their collective identity, register a mix of trimodal responses: some definitely would, some definitely would not, others might or might not. In mode, communitarians would seek something else.

But if they were to seek an academic career again, what would the scientists do differently, again with the knowledge they now possess about science, academia, and careers? Elites claim they would do nothing or very little differently—responses that tend to identify choice and strategy in the selection and pursuit of scientific problems. Pluralists assert that they would remain more focused on and more aggressive about research. Communitarians say they would seek a "better" institution, namely one with a more ready kind of professional opportunity that scientists, at least scientists of this age cohort, have sought.

Correspondingly, in the past ten years, elites have not wanted to leave their universities. Some pluralists have, others have not. In mode, communitarians wish they were elsewhere. Similar patterns are observed regarding retirement—whether scientists would do so now if assured their present standard of living. Elites would not. Some pluralists would, others would not. The same is true for communitarians. Those communitarians who say they would not retire have hope for their academic future or, viewing their age, believe they are too young "to do nothing," but lack a practical alternative.

The characteristics summarized above crystallize into overall modal career patterns found in these scientists at mid-career, described throughout the preceding pages. For elites, this pattern consists in stabilization and rededication; for pluralists, in reversal; and for communitarians, in stasis.

I stress that these characterizations are modal. These characteristics pertain to the majority of the subsamples of scientists of the early- to mid-career cohort. In cases, scientists deviate from these modes. But, as the evidence tells thus far, incidence of such deviation is relatively uncommon. In these subsamples, there has typically been one case that departed from the modal patterns of its group. Where there has been more than one deviating case—in the elite subset of scientists—it was observed how change in individual outlooks and orientations and in career patterns can be brought about by change in institution. In each of the cases where deviation from group modal patterns is observed, the patterns closely parallel the modal characteristics found in the adjacent social worlds. Thus there are "elite communitarians," "elite pluralists," "communitarian pluralists," and the like. The emblematic characteristics of such individuals are presented in other of the columns that codify their traits in tables 21 and 22. The proportion of deviation from the mode found in these subsamples is suggestive of the proportion we may find in populations of scientists who inhabit these prototypical worlds of academe. The women scientists of this cohort (and of the older cohorts), regardless of the academic world in which they are employed, display career patterns that parallel elites.

The data point to patterns that arise and come to differentiate members of a cohort. This begins to prompt key theoretic questions about academic careers: How and why does the observed differentiation occur? What are its consequences for individuals and for institutions? What becomes of academic careers extended further in time? What more precisely accounts for the observed within-group variation? Do such deviating cases persist over time? Or do they come to conform? And why? I shall address these questions in the final chapter. Before attempting to answer them, however, more data is necessary—that from the successive cohorts of scientists—to see what happens to scientists and their careers over other points of time. Thus I now turn to the second of the three cohorts of scientists, those whose passages have brought them from mid- to late career.

CHAPTER THREE

Mid- to Late-Career Passages

The second of the three cohorts under study have passed from middle to late phases of their careers. An examination of this cohort makes it possible to address the theoretic objective that guides the inquiry, namely, how scientists age in relation to their work. In turning to this cohort and the ways in which members account for their career passages, one again confronts ways in which the professional life course in science may be differentiated. On the one hand, it will be apparent how passages through these phases differ among the three main types of institutions where the scientists work—elite, pluralist, and communitarian. On the other, the accounts will begin to reveal, for the first time, how assigned meanings of work, careers, and academe vary between this cohort and its younger counterpart discussed in the previous chapter, as to reveal *cohort differences in occupational aging*, an aim of the overall study.

The organization of this chapter will mirror that of the previous. I will begin by providing a professional profile of the cohort, which will place in context the accounts that scientists provide. I will then summarize the findings obtained when studying these scientists ten years earlier at mid-career, in order to establish a base upon which to build knowledge about career dynamics revealed by the longitudinal work. The chapter will then discuss the patterns gleaned in how elite, pluralist, and communitarian scientists respectively pass from mid- to late career.

PROFESSIONAL PROFILE

I present descriptive professional information on members of the mid- to late-career cohort, comparable to that provided on members of the early- to mid-career cohort discussed in the preceding chapter, in order to place these individuals and their careers in greater context. Table 23 provides a summary of this information. In the table is equivalent information on the

TABLE 23. Cohort characteristics: The mid- to late-career cohort[a]

Characteristics	Elites	Pluralists	Communi-tarians	Overall Average
AGE & EXPERIENCE[b]				
Age Range @ 1st Int.	42–51	42–49	46–55	43.3–52.0
Avg. Age @ 1st Int.	48.0	45.0	52.0	48.3
Age Range @ 10Yr. Fw-Up	52–61	52–59	56–65	53.3–62.0
Avg. Age @ 10Yr. Fw-Up	58.0	55.0	62.0	58.3
Range Yrs. Exp. @ 1st Int.	12–24	9–20	10–24	10.3–23.0
Avg. No. Yrs. Exp. @ 1st Int.	17.0	14.0	18.0	16.3
Range Yrs. Exp. @ 10Yr. Fw-Up	22–34	19–30	20–34	20.3–33.0
Avg. No. Yrs. Exp. @ 10Yr. Fw-Up	27.0	24.0	28.0	26.3
JOURNAL PUBLICATIONS: QUANTITY[c,d]				
Maximum @ 1st Int.	183	152	62	132.3
Minimum @ 1st Int.	30	33	6	23.0
Mean @ 1st Int.	91.3	73.3	30.4	65.0
Maximum @ 10Yr. Fw-Up	330	260	76	301.3
Minimum @ 10Yr. Fw-Up	46	53	6	35.0
Mean @ 10Yr. Fw-Up	148.0	128.0	35.2	135.0
JOURNAL PUBLICATIONS: QUALITY[e,f]				
≥80% published work/1st Int.	100.0	66.6	40.0	67.0
≥70% published work/1st Int.	100.0	100.0	40.0	73.3
≥60% published work/1st Int.	100.0	100.0	60.0	80.0
≥80% published work/10YrFwUp	67.0	66.6	20.0	46.7
≥70% published work/10YrFwUp	83.3	100.0	40.0	66.1
≥60% published work/10YrFwUp	100.0	100.0	40.0	72.0
ROLE PERFORMANCE				
Avg. No. Papers @ 1st Job[g]	15.0	11.3	7.0	11.1
Avg. No. Papers @ Tenure[h]	29.0	24.3	14.4	23.0
Avg. No. Paper @ Full Prof.[i]	52.4	46.0	26.0	41.5
Avg. Time to Tenure (in years)[j]	3.4	4.5	5.2	4.4
Avg. Time to Full Prof. (in years)[k]	7.0	5.3	5.0	5.8

Notes:
[a] Source: Each scientist's curriculum vitae.
[b] Elites: N = 6; Pluralists: N = 5; Communitarians: N = 5. Includes all members of the foundational study sample.

(continues)

TABLE 23. (continued)

[c] The number of publications for each scientist includes all published journal articles, scientific journals being the primary medium through which scientists disseminate their research. The number excludes the following: books; textbooks; book chapters; edited volumes; conference proceedings; invited and contributed papers; book reviews; encyclopedia, world book, and yearbook entries; and articles listed on the individual's curriculum vitae as "submitted," "in press," "accepted for publication," "in preparation," and so on. If the same journal articles was published multiple times (in different venues), it is counted only once.

[d] For foundational study: Elites: $N = 6$; Pluralists: $N = 4$; Communitarians $N = 5$. For longitudinal study: Elites: $N = 6$; Pluralists: $N = 3$; Communitarians: $N = 5$. In both studies, one pluralist is excluded. This case is treated as an outlier because a significant fraction of the scientist's career was spent in industry and thus does not represent an equivalent condition under which the publication records of the other scientists are established.

[e] Percentages of scientists' publications appearing in "major physics journals," those journals to which the physics community assigns greatest value. These include, in alphabetical order, *Astronomical Journal, Astrophysical Journal* (including Supplement Series), *Astrophysical Letters and Communications, Europhysics Letters, Geophysical Research Letters, Icarus, International Journal of Modern Physics* (including series A, B, C, D, E), *Journal de Physique* (including series I, II, III, but excluding IV), *Journal de Physique Lettres* (now incorporated with *Europhysics Letters*), *Journal of Chemical Physics, Journal of Geophysical Research* (including series A, B, C, D, E), *Journal of Mathematical Physics, Journal of Physics* (including series A, B, C, D), *Lettre al Nuovo Cimento* (now incorporated with *Europhysics Letters*), *Nature, Nuclear Physics* (including series A, B), *Physical Review* (including series A, B, C, D, E, L), *Physics Letters* (including series A, B), *Physics of Fluids, Review of Modern Physics, Science*, and *Solid State Communications*.

[f] For foundational study: Elites: $N = 6$; Pluralists: $N = 4$; Communitarians $N = 5$. For longitudinal study: Elites: $N = 6$; Pluralists: $N = 3$; Communitarians: $N = 5$. In both studies, one pluralist is excluded. This case is treated as an outlier because a significant fraction of the scientist's career was spent in industry and thus does not represent an equivalent condition under which the publication records of the other scientists are established.

[g] Average number of journal articles published by the time scientists were appointed to their first job. Other publications excluded. "First job" is defined as appointment as assistant professor or visiting assistant professor. Elites: $N = 5$; Pluralists: $N = 4$; Communitarians: $N = 5$. Cases do not agree with total cohort numbers because of unavailable data. Scientists who began their careers outside of academic science (e.g., as industrial scientists) are excluded.

[h] Average number of journal articles published by the time scientists earned tenure. Other publications excluded. Elites: $N = 5$; Pluralists: $N = 4$; Communitarians: $N = 5$. Cases do not agree with total cohort numbers because of unavailable data. Scientists who began their careers outside of academic science (e.g., as industrial scientists) are excluded.

[i] Average number of journal articles published by the time scientists were promoted to full professor. Other publications excluded. Elites: $N = 5$; Pluralists: $N = 4$; Communitarians: $N = 3$. Cases do not agree with total cohort numbers because of unavailable data or because scientists are not full professors. Scientists who began their careers outside of academic science (e.g., as industrial scientists) are excluded.

[j] Average time in years it took scientists to receive tenure. Elites: $N = 8$; Pluralists: $N = 6$; Communitarians: $N = 7$. Cases do not agree with total cohort numbers because of unavailable data. Scientists who began their careers outside of academic science (e.g., as industrial scientists) are excluded.

[k] Average time in years it took scientists to earn promotion to full professor. Elites: $N = 5$; Pluralists: $N = 4$; Communitarians: $N = 5$. Cases do not agree with total cohort numbers because of unavailable data or because scientists are not full professors. Scientists who began their careers outside of academic science (e.g., as industrial scientists) are excluded.

cohort members' age ranges, average ages, and range of years of experience as a professor, both at the time of the first interview and the ten-year longitudinal study.

When first interviewed, the scientists of this cohort averaged a little over forty-eight years of age and had compiled nearly sixteen and a half years of experience as university professors. Ten years later, in late career, the scientists are, on average, bearing down on the age of sixty and nearing a thirty-year career in academe. The age ranges convey that pluralists of this cohort are slightly younger than their elite and communitarian counterparts, and that some scientists in each of the three sectors are near the mark of a twentieth or twenty-fifth anniversary in academic science. Whether two decades or three, the range of experience the scientists call upon, and the corresponding period in which they may potentially reconsider, revise, or redirect their careers, is substantial.

The quantity of their publications at mid-career was numerous, averaging sixty-five articles across the three institutional types. This average, and the average number of publications in the three sectors in this career period, exceed those of the preceding cohort who have now ascended into mid-career (compare with table 20). This disparity may reflect variation in access to and availability of funding to support scientific research, which was more generous in previous periods, thus benefiting the present cohort, although it is difficult to pinpoint precisely the exact reasons for the variation.

By late career, productivity patterns across the three academic worlds evince two main conclusions. First, stratification in publication quantity holds across the three worlds in a way that corresponds to their prestige and institutional resources. Elites at late career averaged 148 articles, pluralists 128, and communitarians 35.2. Second, the quantity of scientific productivity accelerated between mid- and late career, but especially for elites and pluralists. The maximum output among elites was 330 articles, for pluralists 260. By contrast, communitarians' publication productivity slowed sharply, their maximum of 76 articles more closely approximating the minimum achieved by their elite and pluralist counterparts.

This pattern of limited productivity growth among communitarians was not found among members of the younger cohort (table 20). Instead, publication productivity among communitarians of the early- to mid-career cohort climbed at a rate roughly equivalent to elites and pluralists. This disparity suggests that younger communitarians approached their work with a greater research-mindedness than did their middle cohort peers, even at a time of comparatively scarcer funding for their research, an outlook and orientation to work that was likely fostered by an increasing propensity of institutions of all types to hire faculty with elite doctoral backgrounds, a situation made possible by a glutted doctoral market for faculty positions. Even more noteworthy is that younger communitarians were able to publish at rates higher than their mid- to late-career peers, as their institutions presented considerable challenges, discussed in the previous chapter, for realizing a research career.

The quality of publication productivity also reveals patterns over time across the three academic worlds. At the time of the first interview, elite members of the present cohort, then at mid-career, published all of their work in what the physics community regards as the leading journals. Pluralists approximated these patterns: two-thirds had 80 percent or more of their work published in leading journals; all of them had better than 70 percent of their work published in leading journals. Communitarians published at lower-quality thresholds.

By late career, elites branched out, as it were, publishing less of their work in leading journals. Sixty-seven percent of elites at late career had 80 percent or more of their work appearing in top venues, though all of them published at least 60 percent of their work in what are considered the major outlets. Again, pluralists in these career phases approximated, or placed more of their work in leading journals, compared to elites. Two-thirds of the pluralists had 80 percent or more of their work published in the leading journals; all of them published at least 70 percent of their work in such places. And again, in late career, communitarian patterns diverge. Quality of publication productivity tapers off at the upper-most threshold: 20 percent of the communitarians published 80 percent or more of their work in the leading journals of their field.

At the time of the first interview, all but two members of this cohort were full professors, the exceptions both associates. The exceptions remained associates, one having retired, by the time of the ten-year study. The other scientists advanced in other ways (refer to table 12). Six of them assumed chaired professorships, one of them having achieved this status by the time of the first interview. Another of the scientists retired, bringing the total number of retirees from this cohort to two, both from the communitarian world. The scientists retired from their positions at the respective ages of sixty-five and fifty-six.

Thus, while a couple of scientists of this cohort retired and while a little less than half assumed chaired positions, rank remained essentially the same over the ten-year interim. In ways peculiar to the academic profession, however, rank fails to serve as a significant proxy of career attitudes beyond the point where people cannot advance further in rank, which in science usually occurs, as it does here, in their early to mid-forties, when yet half or more of a career is left to lead. Consequently, one must turn to their accounts for the detailed ways in which the scientists perceive their careers changing and remaining the same, and with what consequences to themselves and to science. Before doing so, it is necessary to become familiar with the patterns that characterized the scientists at mid-career.

MID-CAREER PATTERNS

At mid-career, elite, pluralist, and communitarian scientists had come to perceive themselves and their work differently. In several ways, their outlooks on work and career resembled those of the younger cohort who have now progressed into middle-career phases. In other ways, their outlooks were somewhat different than those now espoused by the cohort directly behind them.

Once again, a series of generalizations may be made about these scientists' mid-career patterns, so as to be able to place their subsequent late-career accounts in greater perspective (Hermanowicz 1998; 2002). The principal findings of the scientists' mid-career patterns are summarized in table 24.

Ten years earlier, elites brought a focus on research to their careers. Pluralists stressed research and teaching, while communitarians had come to stress principally the teaching role in their work. Members of the youngest cohort, however, had come to display a more modulated set of emphases on their career, especially in the pluralist and communitarian worlds (see table 22). Pluralists of the younger cohort, now at mid-career, came to place a heavier emphasis on mentoring and activities related to doctoral training that stand between research and teaching, compared to pluralist members of the present cohort at mid-career. Communitarians of the younger cohort, now at mid-career, came to place a heavier emphasis on service activities and on research, compared to communitarian members of the present cohort at mid-career. In both cohorts, the career among elites is centrally rooted in research.

Professional aspirations of the present cohort at mid-career had also evolved in systematically distinct fashions. Among elites they had intensified; among pluralists they had diminished; and among communitarians they had subsided or were extinguished altogether. These patterns closely

TABLE 24. Mid-career patterns of scientists

Career Dimensions	Elites	Pluralists	Communitarians
CAREER FOCUS			
In Mid-Career	Research	Research & Teaching	Teaching
PROFESSIONAL ASPIRATIONS			
In Mid-Career	Intensify	Diminish	Subside/ Extinguish
RECOGNITION SOUGHT			
In Mid-Career	Great	Average	Minimal
ORIENTATION TO WORK			
In Mid-Career	Moral	Moral or Utilitarian	Utilitarian
WORK/FAMILY FOCUS			
In Mid-Career	Work	Family & Work	Family & Work
ATTRIBUTION OF PLACE			
In Mid-Career	"Haven"	"Happy Medium"	"Stymieing"
OVERALL SATISFACTION			
In Mid-Career	High	Medium	Low

parallel career dynamics observed in the aspirations of the younger cohort now at mid-career. A difference is found among communitarians (compare with table 22). In the present cohort at mid-career, professional aspirations among communitarians subsided or were extinguished, but for their communitarian counterparts now at mid-career, they also subsided, but without cases in which they were wholly extinguished. Again this suggests that the emphasis placed by younger scientists on research across these historical periods left open in their minds the possibility, however small, of significant scientific achievement, so as not to obliterate completely dreams of research accomplishment found in their more senior counterparts.

The recognition that the scientists sought at mid-career evolved in ways parallel to their aspirations. Elites sought great, pluralists average, and communitarians minimal levels of recognition. A difference is found among elites of the current and younger cohorts. Younger elites now at mid-career came to seek substantial recognition, a more muted level of reward that may owe itself to the heightened difficulty of securing funding necessary for achievement and its associated recognition.

Elites at mid-career were morally oriented to their work: science was an end in itself. Pluralists, and especially communitarians, had developed a more utilitarian orientation to their work: it had become a means to the end of earning a living. These patterns in orientation are identical to members of the younger cohort now at mid-career.

Patterns are similar between cohorts and across the academic worlds in the assigned relationship between work and family. Elites stressed work, pluralists and communitarians family and work. Pluralist and communitarian members of the younger cohort appeared to assign more priority to family at mid-career than did their more senior counterparts in the same phase.

How did the scientists regard the departments and universities where they worked? Elites regarded them as a "haven." Elites of the younger cohort now at mid-career, recall, attributed them to be "the best," a similar characterization, the former conveying a sense of settlement, the latter a newly realized condition. Pluralists of the present cohort at mid-career found their departments and universities to represent a "happy medium," much as their younger counterparts did at the *start* of their careers, where research and teaching roles defined daily life but without what they would deem as excessive expectation. Pluralist members of the present cohort did not view their scientific homes as "dens of confusion" as their younger counterparts did by the time they reached mid-career phases, a disparity that likely has its roots in the heightened competition for funding, which when combined with the comparative latitude in pluralists' roles, produces the disaffection evident in contemporary pluralist scientists at mid-career. Communitarians found their academic worlds "stymieing," deadening for the research they had aspired to conduct. Their younger counterparts now at mid-career evolved from this stance to that of seeing their institutional world as "a job," a less negative attribution that again likely owes itself to the research push and promise, however faint, that remains felt in the minds and mores of contemporary communitarian scientists at mid-career.

Overall, elites were highly satisfied in their careers at middle phases. In comparative terms, pluralists assumed a middle level and communitarians a low level of satisfaction. Between the cohorts, pluralists changed the most in this regard. Contemporary pluralists at mid-career evinced comparatively low overall satisfaction in their careers, a culmination of the confusion they had begun to discover about their roles. They sought answers to the question of which of their roles to emphasize and which to downplay; which roles to seek and which to leave behind.

What has become of the scientists, who ten years previously were at mid-career? How do they now view themselves and their work, and how do their perspectives vary by academic world? It is to these concerns that I now turn.

ELITES

In following elites from middle to late phases of their careers, one encounters relative continuity, a chief finding about scientists in this phase in this world of science. Findings of dramatic change make for dramatic stories of people and their careers. There are such stories, but not now and not here. Instead, the modal career pattern is *continuation*. Elites in late career often see a diversification of their roles compared with prior times; many enter posts concerned with the administration of science or of academia generally or find their customary role triad more laden with administrative duty. But overall, scientists see and speak of an extension of that which defines their primary roles as research scientists, which, with rare exception, is construed by them always to stand above all else.

It is, then, common for these scientists to furnish the seemingly mundane responses of, "There's been no change," or, "I don't expect any dramatic change; I have just carried on with my work," or, "The last ten years have been relatively uneventful . . . just the standard progression," to characterize what, in fact, turns out to be the standard. In keeping with the refrain, elites are "busy." "I'm so busy," they say. Everyone is "busy." No one has "enough time" to do it all. "I feel like I have much more to do than I had before. I feel like I'm a lot busier than I was, whereas I expected the reverse" (18L). "I'm just too busy. These other responsibilities, and I have small children, so I have less time at home. It's pretty hard to have an administrative position, children that you take care of, and do a lot of research. Don't have time [for it all]" (6L). "[I am] more busy, equally productive, but the advantage gained through experience is being used to the fullest to maintain productivity" (17L). "I'm busier. I'm busier. I'm tired of having to write grants, I have to admit to that. That's for sure. I love doing my research work, but I'm really tired of having to write grants and having to go through all the effort that you do to try and keep money coming in for research. . . . You have to compete" (18L). Finally, "I'm more tired. That's seriously the main thing that comes to mind. I'm less able to have time for contemplation, including professional contemplation. That's a problem. . . . It really can be quite tiring and overwhelming at times. It's very exciting intellectually. There's a lot of new things going on. . . . In a position like this, you just don't have enough time to do everything you'd like to do" (19L).

Along these lines, elites in late career are most apt to report that they work as hard or harder than ten years ago, that they are equally ambitious for scientific achievement, and that, overall, their careers have progressed as expected, if not better than expected. Like their younger counterparts, their overall satisfaction with work is comparatively high. Despite gripes about time, competition, and funding—the latter being their principal career concern—their work attitudes are positive, and they remain, much like their younger counterparts, morally oriented to their work: science for these scientists is its own end.

For these reasons, the idea of ever leaving science—an idea that these scientists come to "bat around" as they enter phases of the late career—prompts notable ambivalence. Clearly, it is easier in the minds of these elites to picture continuation in their roles than removal from them, even in anticipation of the most advanced phases.

> I think probably now I tend to think more, or I have thought a couple of times about, okay, I'm now sixty-one years old, and people tend to work to maybe seventy, although my wife tells me probably I shouldn't retire, because I would keep working anyway. And I might as well get paid for it. So I don't really have any definite plans to retire yet, and I sort of figured, as long as it's still fun, and I probably find more to do, [why retire?] . . . I don't think about really doing it because I still really enjoy the work and the setting.
>
> INTERVIEWER: Do you envision ever retiring?
> SCIENTIST: Logically, yes, but I don't see what it would be. . . . If I became physically incapacitated and clearly couldn't do what I was supposed to do, I could see the possibility. (17L)

> INTERVIEWER: If you could retire now and lead approximately the same quality of life, would you?

SCIENTIST: No, because my quality of life is all bound up in what I do.

INTERVIEWER: And you couldn't leave this?

SCIENTIST: No. It's where all the intellectual meat is. . . . It's too bound up in what I do. (12L)

Echoing these sentiments, another of the scientists decoded the institutional ritual of retirement, at least as it informs the mores of this academic world: "Many people, in my experience, equate retirement with death, and I think that's why sometimes scientists don't retire. When I retire, I will keep contact with physics. I'll still think about physics. . . . [But,] I'm not ready" (19L). Or, as another scientist put it: "I'm not sure it's possible to do what you've just said. To retire would mean not to be involved in research work, so the quality of life changes. It's a different quality, let's put it that way. I don't think I would want to retire in the sense of stopping research work" (18L). Or further still, as one of the scientists put it most simply: "I might retire. But only if I can still come in and work" (7L).

Continuity of career patterns extends to other defining characteristics for scientists in late phases, in several ways making them indistinguishable from elite scientists at mid-career as discussed in the previous chapter. They remark, with virtually no variability, that the point of their peak career satisfaction is the present. Consequently, not only would they not retire if given the chance, which, for the most part, is a chance they now have, but they also assert they would do nothing differently in their careers, nor leave their present universities, which they regard as a haven for the articulation and satisfaction of their professional and personal interests. Professional and personal interests often seem indistinguishable.

It keeps me very satisfied. You're your own boss. You decide what you want to do. That, to me, is the most important thing.

INTERVIEWER: Would you do anything differently, knowing what you know now?

SCIENTIST: No. Somehow I made all the right choices without knowing it.

INTERVIEWER: If you were to go home and complain about something, what would it be?

SCIENTIST: It may sound strange, but I don't have many complaints.

INTERVIEWER: Are there any ways in which you have found an academic career unrewarding?
SCIENTIST: No. I'm pretty happy with where I am. (9L)

INTERVIEWER: Has there been a period of your career or age when it's been the most satisfying?
SCIENTIST: Now is probably pretty close. But, again, I've been at a high plateau for a long time. . . .
INTERVIEWER: If you were starting all over again, knowing what you know now about academic careers, would you go into one?
SCIENTIST: Oh, yes, absolutely. I have no regrets whatsoever.
INTERVIEWER: What would be the greatest draw?
SCIENTIST: Scientifically you get to do something you really enjoy, and in some way I've felt that people are paying me to play with very expensive toys. . . . I enjoy interacting with a lot of very smart, very diverse people. I enjoy travel at some level. It can be too much, but I've seen a lot of the world that I never thought I would see. There are many attractions.
INTERVIEWER: Would you do anything differently, knowing what you know now?
SCIENTIST: Not much. I can't think of any decisions that I would make differently, looking back on it. No, I really don't. In detail, would I have done something differently that day, did I accidentally say something that I now realize hurt someone else? Yeah, sure, we all do that, but that's not what you're asking.
INTERVIEWER: What are the major complaints you have about your job at this time?
SCIENTIST: Not enough hours in the day, and no one can solve that. That's really the only complaint. I get great support from the president, from the provost. We have a new president, and I think it's working out very well. The previous president also was terrific. This is a great place. I'm very lucky. (19L)

While elites in late career are morally oriented and exhibit positive attitudes toward their work, one is able to detect some evolution in their professional aspirations and in the way they seek recognition from professional peers. As with nearly all scientists, indeed as with other occupations, aspirations become more finely calibrated, so as to bring people's expectations for achievement in greater concert with both structural opportunity and one's own individual ability to satisfy them. This general process is found throughout academic worlds, including the elite, even where mandates for achievement remain strong throughout a career, as evident in the productivity patterns across the three academic worlds. While becoming more realistic about expectations for achievement is a phenomenon that grips elites too, nevertheless those expectations remain pronounced in the elite world. The evolution, then, essentially consists in proceeding from the loftiest of achievement expectations to the merely lofty. These scientists realize they may not be the next Newton, as they usually once dreamed from adolescence to their first years in academe, when realities of academic work, particularly its ardor, are made most plain by the first full-fledged onset of professional independence. Newton and the others who comprise the scientific pantheon notwithstanding, these scientists still seek late in their careers highly substantial scientific accomplishment. The level sought reminds them of how far they have left to go, but despite the distance, the level remains real in their minds. Following W. I. Thomas's classic theorem, it remains real in its consequences (Thomas 1923). This is a world that propagates high expectations and a world that sustains those expectations because a rare few actually satisfy them.

> I know that I am just as ambitious about my research. . . . I have the same longing to succeed. I'm a little more tired and a little more distracted. I don't think I have accomplished what I wanted. I haven't really done something of everlasting importance, and so I'm very disappointed in some aspects of my career. I want the satisfaction. I don't want the glory, I want the satisfaction. I work in an environment where I look at some of the things that people have done, and they really made serious, major contributions to science. I don't feel that way about myself. That's a little disappointing. When I was younger, I was more optimistic about doing that. I'm 52; it's harder to believe it can happen. My grasp is greater than my reach. I always want a little more than you can have. . . . The guy down the hall wins the Nobel Prize—it just happened, two doors down. It's kind of oppressive in a way. There is always a fuss, and reporters come, and the TV cameras march in. There's all this attention going to the guy in the next office. I can't really complain about that, but sometimes that is something you wouldn't mind skipping.

INTERVIEWER: On a scale of one to ten, ten being greatest, how satisfied with your career would you say you are?

SCIENTIST: I'm pretty satisfied. I would say I'm probably an "eight." . . . and not "ten" because I haven't accomplished some things I really would like to accomplish, to do something more significant, to do something that matters in a deep way, something discovered that is fundamental and really important. Not just writing a paper, but writing something with really lasting value. I've written a lot of papers; I'd be dishonest to pretend otherwise. But I haven't written anything that's really of lasting value, that really changes things. . . . My aspiration now is to try to solve some important problem. I do have dreams, [and if the dreams come true], we'd have a profound effect on computer science. We'd change computer science. It's for the high stakes. (6L)

I would say "eight" [on the scale of satisfaction]. There's no perfection, right? I can certainly imagine getting a Nobel Prize tomorrow. Then it would be a "ten." But I haven't gotten the Nobel Prize. I have to leave room. I would say "eight" or "nine." There's always room for improvement. There are problems I would certainly like to solve, certain problems I'm struggling with. . . . There are problems I would like to solve in this field. There are big problems I've been working on for ten years. (7L)

Probably "nine and a half" [on the scale of satisfaction]. [I would like] more personal recognition from the outside. It's not like I haven't gotten any. I have a chair, I've gotten teaching prizes, I've gotten research prizes. . . . Since the last time we spoke, one of my former students got a Nobel Prize in physics. It was great. My wife and I went to Sweden [for the award ceremony]. It was a wonderful thing. But it's just . . . I don't know how to explain it exactly. It's not enough. It would be nice to have a little more. . . . I think I am more satisfied now than I have ever been. . . . I would very much like to be able to have more time to just sit and think, just sit and think about something, rather than be pressured to write a [grant] proposal, pressured to write a [grant] progress report, pressured to do any of that. . . . I have always felt right from the beginning, thirty years ago, of having a vision, which is a dream of using theory to really do some great things. My dreams have broadened in scope . . . [it's] very, very hard [to accomplish] (18L)

In these accounts and in others like them, the source from which scientists define success in their work and career is plainly evident: it is external, bound in the norms that govern scientists in their roles as researchers. In forming estimations of themselves, they turn not simply to colleagues down the hall, but to those corridor colleagues who are in turn well integrated into the institutional goals of science. Direct comparisons to establish self-worth may at times be local, but the gold standard upon which any comparison achieves meaning is cosmopolitan (cf. Gouldner 1957–1958). This was found among elites at early and middle phases of their careers, and it remains unchanged in late phases, proving to serve as one of the chief continuities situating and defining careers in the elite academic world.

An embrace of the scientific reward system is associated with these patterns. It is the mechanism by which recognition is granted or withheld, using external standards of the wider community to prompt or deny rewards, regardless of whether the rewards flow from local departments or more nebulously defined scientific communities outside of one's immediate institution. While in general cognizant of what they take to be its imperfections, elites find the system fair overall. Whether their view of the reward system would be so favorable were they not major beneficiaries of it is an open question, and one that can be answered partially through comparison with scientists' sentiments in other academic worlds. Elites account for the reward system in a manner typical of the following scientists:

> The reward is important. I am comfortable with what I have achieved. I am constantly surprised that I have actually accomplished as much as I have, as much recognition. . . . The Buckley Prize I liked very much. That's the biggest satisfaction [beyond election to the National Academy], the recognition by your colleagues. The Academy was very nice. It opens doors. You catch people's attention.
>
> INTERVIEWER: What do you think of the many scientists who would very much like to become members of the Academy but have not been inducted?
>
> SCIENTIST: There are many of them. Somehow it's something people want. It's a good goal. I mean it's nice to have this recognition. . . . The whole election process now is fairly political. I wouldn't say it's a very good marker. I mean the people obviously belong there, but there's a large gray area. [The election process] goes through many, many votes and so on. There are discussions. It's very subjective. The members are all elderly and somewhat out of touch. (7L)

INTERVIEWER: Do you think the system of recognition in science is fair?

SCIENTIST: Yes, I think it is, because a lot of it involves reviews of papers, refereeing, which is anonymous. I'm not saying that people have the best judgment, but I think the system, as a system, is the right way to go, and we learn to work with it.

INTERVIEWER: Have you felt that there have been any instances when your work has not received the kind of recognition it was due?

SCIENTIST: Yes, all the time. That happens all the time, especially if you have a really creative idea. But in the end, I realize it's the person who is writing the paper, and it's our fault for not getting the message across. We can't get the message across. Somehow we did something that made this person reading this think they didn't get the complete message. When they do finally get the message, I believe it does matter. I think a lot of the responsibility is us making sure we make a good case. Very often what will also happen is that people will just not see it the way you do. We live with that. Eventually, things get recognized. Good science eventually gets recognized one way or the other.

INTERVIEWER: Are you saying that recognition is often delayed?

SCIENTIST: Yes. And that's a function of several things. It's a function of how novel the idea is. The more novel the idea, the longer time it takes. It's whether it's a theoretical idea or an experimental idea. A theoretical idea takes a very long time to be appreciated because someone needs to do an experiment to verify it. An experimental idea usually gets recognition much faster. (18L)

Thus in elites' eyes, the system works despite its "slights," which, in any event, may be as much attributable to individual failings as to systemic ones, upholding the merit-based ideology on which the system relies.

When first interviewed, one member of this world and cohort deviated from the predominant patterns in ways akin to communitarians: the scientist largely withdrew from research and embraced teaching in mid-career. Ten years earlier, the career was understood in these terms:

> I am doing what I like to do differently. I'm teaching more. I'm doing more administration.... I've been teaching a large freshman physics class for the last three years.... It has seven hundred students; it's by far the largest class [at this institution]. I made the choice clearly to

put that kind of time into something. . . . I think there has been a very fundamental personal change in what I think is important. It used to be that physics really was central to my life, whereas now I don't feel that way. That's why I'm teaching. . . . I have changed focus in terms of what I consider important. . . . I have the feeling that publishing papers in a narrow discipline—I'm not very interested in that. It just doesn't seem like a worthwhile place to put my effort. . . . I'm not drawn to research because it seems very dry, and at this stage of my life it doesn't seem particularly rewarding. . . . Teaching [freshmen] well is something I feel is purposeful. (12F)

What happens to this type of scientist ten years later? Is he brought back into the fold, as it were, to conform to performance expectations indicative of the roles in the elite world of science?

The last five years I've been doing almost no scientific research. . . . I've done what I wanted to do. I've been successful at doing it: I've raised five million dollars all total to change the way freshmen physics is taught here. A couple of weeks ago, there was an article in [a national newspaper], where what we're doing was featured very prominently.
INTERVIEWER: . . . Are there ways in which you feel you have been unsuccessful?
SCIENTIST: I'm sure this educational thing is going to continue on. It has been a lot of effort. . . . A [non-faculty] staff physicist [at this institution] said to me, "You know what happens with physics researchers who get old? They start doing physics education research, when they can't do real physics research anymore." I think I've been as successful as my talents justify. I haven't gotten a Nobel Prize in physics. When I was younger, I thought about that. But I've done a lot of [other] things. (12L)

Ten years later, in late career, the patterns not only remain, but have deepened. In this sense, there is further development of—and overall continuity in—deviation. This trend shall be important to remember. Accounts will reveal other examples of "deviating cases" in the consideration of other worlds and other cohorts and will thus present the occasion to see how the cases change, or remain the same, over time.

PLURALISTS

Pluralists' passages into late career are bimodal. One mode is marked by less continuity in their professional roles and outlooks compared with elites. The other mode resembles that of elites. Owing to their collective identity, pluralists exhibit this broader variety of career patterns than found among elites or communitarians: they are a mixture of them. Recall that, while at mid-career, these pluralists found their world a "happy medium," they also saw their loftiest aspirations diminish, largely because of incongruent estimations between the achievements they sought and the opportunity in their academic worlds to foster such achievement. How would they proceed?

In one modal pattern, pluralists in late phases of their careers have proceeded by attempting to *regenerate* themselves as scientists. They seek to find professional activities, if possible in research, to sustain professional involvement. In the other modal pattern, pluralists *continue*, like their elite counterparts: they identify strongly with and are successful in research and seek to conquer new research frontiers, this despite an institutional culture and structural set of resources that can blunt such aspirations, as was evident in many of these scientists at mid-career.

The two modes are associated with a split of career patterns. Pluralists in regeneration identify more with teaching, pluralists exemplifying continuation, with research. In regeneration, pluralists have seen their professional aspirations subside further, while, in continuation, pluralists' aspirations have generally remained steady. In regeneration, pluralists are marked by a deepened utilitarian outlook on work, although they are questioning the viability of this outlook, whereas, within continuation, pluralists remain morally oriented to science. In regeneration, family has assumed greater prominence in the constellation of scientists' roles; in continuation, work remains a "central life interest" (cf. Dubin 1992).

For those who embrace teaching more strongly, for whom family has assumed increasing prominence, and who generally espouse a utilitarian orientation to their work, the career is often said not to have gone as expected. Such scientists also tend to identify their peak point of satisfaction as early career when, as it turns out, they were more morally committed to science. They question whether they would again seek an academic career, regard the scientific reward system as unfair, and, correspondingly, have adopted a more internal, as opposed to external, means of defining success.

By contrast, for those who continue to embrace research, for whom work has remained central — "above" family as it were — and who generally espouse a moral orientation to their work, the career is typically said to have gone better than expected. Such scientists tend to identify their peak point of satisfaction as the present, would readily seek an academic career again, regard the scientific reward system as fair, and, correspondingly, continue to work in relation to an externally situated definition of success.

The first of these modes — the one associated with embracing the teaching role and in which scientists seek to regenerate themselves — is illustrated by the following scientist, who at age fifty-three has served on his university's faculty for twenty-three years, the last twelve as a full professor.

>   INTERVIEWER: How has your outlook on career changed or evolved?
>
>   SCIENTIST: Well, it certainly has . . . I think about when I will retire. . . . I think about what I want to do — do I want to stay and work at the university or retire or go do something else? Those were things I certainly wasn't thinking about ten years ago. . . . I'm in the process of trying to figure out what I want to do and how to go about doing it. An aspect that comes into play is that funding for research, funding for higher education, is going down. It makes it much harder to keep things going. . . . It's a little discouraging to put a lot of time into thinking out the details of some worthwhile project, getting excited about it, and not having it there. It's discouraging. . . . I am more involved with things outside of the physics . . . than I was ten years ago . . . largely it involves [my] two kids. I coach soccer teams, I attend soccer practices, my kids started piano lessons about six years ago, so I did too. As you approach fifty . . . you begin to think about all of the things you put off, thinking, "Oh, I'll get to that one day." Then you decide you better start doing some of them. . . . I'm less involved in research than I was. That puts me in the situation of thinking, "Do I want to get more heavily involved in research when the kids get a little older?" I have to think what I want to do with my time because it takes a lot of work.
>
>   INTERVIEWER: What are the possibilities?
>
>   SCIENTIST: The possibilities are to come up with more research ideas, submit more proposals, and eventually get to where I have several projects going on instead of basically one project now. I am not interested in going into administration, I just don't think that suits my temperament. . . . I haven't figured out all the possibilities. . . . Would I advise my children to go into academic physics, if they were doing

it now? I certainly wouldn't push them.... I like teaching, I like the environment, I like being around students, I really like that aspect of it. The question is, if I went into it [again], would I even get tenure? I might. But it's not like I have to do this or die, that's for sure.... It gets discouraging when you go through periods when you don't have any funding, which is going to happen to almost everybody, at least in my field. That's hard.

INTERVIEWER: ... Do you define success differently now than ten years ago?

SCIENTIST: ... I find myself less worried about what my peers think and more interested in trying to figure out what I think. What success is, is a little nebulous right now.

INTERVIEWER: ... On a scale of one to ten, how satisfied would you say you are in your career?

SCIENTIST: "Five." ... You know, research isn't that satisfying. The students that I've been getting lately haven't been that good. I write proposals. Sometimes they're funded, sometimes they're not.... I've always liked the teaching, I really enjoy that. I now run a program, we have a summer research program where we bring in undergraduates from around the country ... and I run that program.... I enjoy that, I enjoy the teaching.

INTERVIEWER: ... What would you say are your current aspirations?

SCIENTIST: Well, that's what I'm trying to figure out. I know you have to ask that question. That's my aspiration, to figure out what it is I want to do five and ten years from now.

INTERVIEWER: Would you say that you have a "dream," something you want to attain?

SCIENTIST: No, but I should.

INTERVIEWER: ... Do you have worries or concerns about your career?

SCIENTIST: My worry is that I'm not sure what I'm going to be doing five years from now.... I don't know whether I'm going to fire up and spend ten years after that working in a lab. Do I really want to fire up and spend a lot of time in the lab? I worry about resolving that issue.

INTERVIEWER: Does it keep you up at night?

SCIENTIST: Yes. It does more than the research funding [concern]. I worry about making this decision. I worry that I should know what I want to do right now and I should be working toward it. I feel like the day is going by and think, "Why isn't this obvious?" I should know what I want to do. (48L)

Ten years earlier, the scientist may have had answers to these questions, as the follow-up evidence leads one to believe. A decade earlier, at age forty-three and two years into a full professorship—a point that some might regard as a pinnacle in the academic career—the scientist accounted for his career in the pluralist world in this way:

I'm at a point of figuring out what it is I would like to do, if there is anything different I would want to do in the next twenty years. I'm on a certain track now, and I'm trying to see where that is going to take me and what other options I have. I wouldn't say I have aspirations to be an administrator or to go work at the NSF [National Science Foundation] or to win a certain prize. As far as what is the next plateau to reach, I don't know.

INTERVIEWER: Is there some ultimate thing you would like to achieve?

SCIENTIST: No. There's not a specific thing. My oldest child is almost five. A lot of my focus and aspirations have been diverted to having a family. . . . I spend less time [with my work]. I spent almost all of my time doing it before. (48F)

While the scientist has grown increasingly ambivalent about the academic career, the comparative lack of satisfaction derived from research and the funding process, it is revealing to know what this scientist, as with other pluralists like him, would do differently were he to seek an academic career again, knowing what he now knows. It is revealing because it identifies the strength of the institutional norms of science that, in these cases, attempt to pull scientists back in the grip of research, even as they come to identify more with their instructional roles. For this and for other such scientists, this desired knowledge amounts to the maximization of information about that which is necessary to thrive as a researcher:

I would have maintained closer ties with my senior colleagues because I needed to manage my career better. I needed somebody to talk to about how much time I should spend writing papers, how much time writing proposals, how to pick graduate students.

INTERVIEWER: Have these had costs for you?

SCIENTIST: Yes, I have made a lot of mistakes. It would have been nice to, in effect, have a manager who could make sure that I was writing proposals regularly and thinking about research directions. (48L)

The response is instructive on two counts. It illuminates, on the one hand, scientists' sense that steadiness in research brings about steadiness to the career, a pattern observed previously among elites: research appears to institutionalize career continuity as well as identification with the career. On the other hand, the response provides ample indication of the pluralist culture: while scientists may embrace a plurality of roles, sometimes stressing one over the other in various career phases, this can entail significant costs, at least as evidence suggests here, when teaching assumes a position superior to that of research. An academic environment that is so flexible, creating multiple and competing systems of reward, appears to introduce career risks to individuals, namely those that jeopardize the consistency of their role commitments and research productivity. Given the nature of what lies in jeopardy, these are as much risks to individual scientists as they are to the institution of science.

The patterns described above create ambivalence toward the scientific reward system.

INTERVIEWER: Do you think the scientific community, and its system of recognition, has acknowledged your contributions fairly?
SCIENTIST: No. I think we have done significant things, and if they don't get recognized, it's discouraging not to get recognition for those things.
INTERVIEWER: For research?
SCIENTIST: Yes, for research. Teaching is its own reward.
INTERVIEWER: Are those factors strong enough to make you think twice about going into an academic career again?
SCIENTIST: Yes, those and others.
INTERVIEWER: The others being?
SCIENTIST: The uncertainty of getting funded, the uncertainty of good graduate students. (48L)

Another of the pluralists remarked similarly:

[The reward system is] very far from perfect. The true geniuses will be recognized regardless; they'll be rewarded and recognized. But after that, it depends very much on the administration [of science], or how it's done. It is the "loud people" who get more recognition. That's not necessarily according to merit. . . . Being aggressive and loud, it seems to pay. . . . To be honest, unless a person really wants to do it, I wouldn't encourage a person to go into [an academic career]. (52L)

Nevertheless, scientists who exemplify these patterns claim not to have wanted to leave their institutions, a finding that likely speaks more about the opportunity cost of leaving than an actual desire to stay. The sunk costs of scientists in these phases of their careers well exceed realistic opportunities for alternative employment. Still, they would entertain the possibility of retirement, if they found a worthwhile alternative to their current way of life, although this, too, is seen, in the end, as too costly in practicality.

In other pluralists, the contrasting set of patterns is evident, those in which the mode follows *continuation*, akin to the elites of this cohort. The patterns are revealed in the following accounts.

I like to work, and I like solving puzzles. I'm very competitive. I remain competitive. The fields continue to evolve; we continue to invent new fields. I like being out there and presenting them, and I like looking at possible inventions that become commercialized, which is part of what we [in this lab] do. . . . [I'm] harried, because there are more things that come at me . . . the requests for my time have gotten worse. . . . I certainly can't speed up much. I've already worked my seventy-hour week. (58L)

Like elites of his cohort, the scientist exemplifies an institutional press not only for continued scientific achievement but for achievement of a particularly high order.

We haven't gotten the Nobel Prize yet. That's something that is proposed occasionally. We did do work [in that vein] with the initial discovery of this one particular field.
INTERVIEWER: On a scale of one to ten . . . how satisfied in your career would you say you are?
SCIENTIST: Probably "eight," "nine." . . . I haven't gotten the prize yet, but I'm not making it the end reason. [I would like] something over the top, and a legacy to the field. I look at the thirty-six students who will receive their PhDs in my twenty years. That's an extraordinary

number. Of the thirty-six, thirty-five are successes—that's a great percentage.

INTERVIEWER: Is there a period of your career or age [when] you have been the most satisfied?

SCIENTIST: Now. I've accomplished a lot. But I want to accomplish a type of satisfaction of having solved the major problems [of my field]. . . . I'd like to achieve some very significant things. I'm interested in creating new areas and establishing [ideas and work] that are strong, that have an impact that is broad. That's my biggest agenda.

INTERVIEWER: . . . If you could retire now and have the same quality of life, would you?

SCIENTIST: No. I'd probably be bored. I'm used to being in this challenge of discovery and making things happen. What I see more likely is simply not retiring. (58L)

Another of the pluralists accounts for his career in similar terms, highlighting the patterns that comprise this modal passage:

I've gotten more successful overall [compared to ten years ago]. Things are much more hectic. I have too many things to do.

INTERVIEWER: In what ways have you become more successful?

SCIENTIST: I have much more money to spend. I have many more students and postdocs. I have a lot of international collaboration.

INTERVIEWER: Have you seen this change in the past decade?

SCIENTIST: Yes, yes. . . . I enjoy what I'm doing, but I realize I'm almost sixty years old, and it's going. I'm going to have to start slowing down. But right now things are at a peak.

INTERVIEWER: Would you say you are at your peak now?

SCIENTIST: Yes. It's very late in science. . . . I can't conceive working harder. There isn't enough time in the day. . . . I think I've come closer to what I thought was success. I'm in a good position. . . . I'm where I always wanted to be, at least with the size or the group doing the amount of projects. I still sometimes feel I would have been happier at a more powerful place, but I'm not sure that's true. . . . I think I could do more at a better place. I think we all feel this way at places. The better places—that we belong there, and I can't get rid of that [feeling] because I have been offered positions abroad comparable to the best places here. So that's bothered me. It doesn't happen in my own country. I'm bothered by the fact that I can win these foreign awards and never win anything comparable here. I think that's because I come from [this institution], and no matter what I do, [that will not change].

INTERVIEWER: ... Do you feel sufficiently recognized for your work?

SCIENTIST: Oh, yes. Yes. It's just I would like some of it to be a little closer to home, but not as close as the local university. I think it's just more international than national. That just makes me crazy. So it would be nicer. I would like to become a member of the National Academy of Sciences. But again, it's something that's not likely. To give you an example about this institution, several people from midwestern universities have won the Nobel Prize without being members of the National Academy, until afterwards. It's a very closed organization. That would be something that would be nice.

INTERVIEWER: If you could retire now, and lead approximately the same quality of life, would you?

SCIENTIST: I don't think so. I don't think so. No. I think what you mean by *retirement* is leave the field. It's not what people typically do when they retire. They don't teach, but the research goes on even faster. For example, the guy I collaborate with at [the University of] Chicago is retired, in a sense. He goes to his office everyday at seven o'clock and starts working. He's retired. (47L)

Finally, the third of the four women scientists in the sample, and the only woman scientist of this cohort:

I would [go into an academic career again]. I'm interested in physics and want to know answers to questions, and an academic career allows you to spend a large fraction of your time trying to answer those questions. I don't know any other company that would do that.

INTERVIEWER: ... Are there ways in which you have found an academic career unrewarding?

SCIENTIST: No, not really. I suppose that's why I'm complacent. I'm satisfied. (52L)

It is noteworthy that many of the pluralists who exemplify patterns of continuation (the woman scientist an exception) began their science careers elsewhere, in industry or in other universities supportive of research. Those who exemplify the regeneration patterns began their careers at the pluralist institution under study, having effectively been "raised," and socialized to career norms that pervade this academic world. By way of accounting for his career and attempting to explain his orientation to work vis-à-vis other scientists around him, one of the pluralists—who falls into the continuation mode—made several observations about the nature of institutional culture that appear to mark the pluralist world. In particular, he was drawn to comment critically on the multiple systems of reward found commonly inside pluralist (and communitarian) institutions; these systems typically function to pit teaching and research against one another, the result of which may often be a weakening of a "research culture" and attendant identifications with sustained research effort in a career.

> I don't think there's much of a research culture as there is at [the University of] Chicago [for example, where the scientist has a main collaborator]. I sometimes think that the administration here rewards research, so that maybe some people become a "distinguished professor." Other times, I think they don't. For example, the department chairman has given me new courses to teach each of the last four terms, making my life much more difficult. It just strikes me that maybe he's jealous. I don't know what the reason is.... There's no stimulation [here] at all. One of the things I've not found here is stimulation from faculty. So to counteract that, I've become a journal editor. It lets me keep track of what other people are doing. I see everything in my field that's submitted. Secondly, I get out of all local committees. So it's very nice. I recommend it to everybody. It can be aggravating, but I think in balance, a very good thing is that it helps the weakest aspect about being here. That is, there is no stimulation from other faculty.... Teaching can pull people out of research. To some people, that is important. It strikes me that some of our best people are teaching first-year courses at levels that would not exist at the University of Chicago or Duke or Harvard. They strike me as enjoying high school teaching. But I think high school teachers are better trained at teaching that sort of thing than these people.
>
> INTERVIEWER: What are the incentives to do that?
>
> SCIENTIST: We have teaching awards that give money [to the recipients], and our department seems to be very good at winning these awards. [The research culture] is handicapped. The thing that happens is that people seem to give up. I have not given up because I did not

come here from the start of my career. I'd say 10 percent of the people don't give up at all. (47L)

By one view, academic institutions with such an arrangement of competing reward systems introduce a means by which academics can age in their careers. When confronted with comparative failure in research, and if, as in these instances, academic positions are secure by tenure, academics can turn to alternatives that institutionally sanction them as "meaningful contributors." The alternatives serve to sustain a valued means of membership in a community, when the most vaunted form of activity — research — has foundered.

By another view, institutions with such arrangements sow the seeds for these types of outcomes. More precisely, they foster muted research achievement and thus legitimize systems that run counter to the institutional goals of science. The reward structures exist when individuals enter the institutions, and they inspire a specific cultural logic. Individuals are thus socialized into a particular institutional milieu that allows, and indeed may even encourage, role plurality. The academic profession, or at least some worlds of it, has adopted the label *scholar-teacher* to ritually designate this academic breed.

COMMUNITARIANS

Communitarians fortified a retreat from science in their passage from mid- to late career. As noted earlier, two members of this subsample had in fact retired, at the ages of sixty-five and fifty-six, by the second time I interviewed them. Others suggested that, insofar as their work, they were "putting in time" in order to retire, now centrally on their minds. While many continue to find teaching often the only source of satisfaction in their work, this, too, frequently connotes working out of necessity rather than enjoyment. Their patterns in late career suggest substantial disengagement, and often disillusionment, with academia. They typically voice severe criticism of their institutions, in part because they understand them as the organizational bodies that have substantially structured and, in their eyes, limited, their careers. In mode, the constellation of these patterns may be best characterized as *demise*, a period in which communitarians resign themselves — now without much resistance — to situations of constrained opportunity and curtailed ambition.

The scientist whose sentiments are conveyed below is typical of outlooks adopted by other communitarians of this cohort.

I don't really care. I'm sixty-five right now. I don't really care about becoming, let's say, a full professor, or any formal steps in my career. Also, recently, I don't have too many graduate students, and essentially none at present . . . the quality of graduate students is decreasing. . . . [I ask myself,] "What's the reason to add one mediocre paper to fifty-plus existing ones [that I've written]?" There's some amount of disillusionment and some effect of being bored and tired. . . . I'm maybe more introverted and less interested in formal aspects of science, career-making, and formal recognition. . . . The last five years has been a slowdown. I could have done more. When I look back into the last five years, I'm not so pleased with myself, indeed. I could have done more. I could have done more. . . . Right now, I'm thinking that this is the best way of graceful decay. (24L)

Another of the scientists viewed the evolution and adaptations of his career in another common fashion:

When I first started out, I would have liked to have become a famous astrophysicist. In fact, I worked very hard at it. I came up with a creative research program, a very creative research program that nobody ever thought about before. While I was here for my first ten years, or fifteen years, I pursued the research and wrote a dozen papers or more on the research. . . . And I always came up a little bit short of where I needed to be. And that was disappointing. That was disappointing. I always said I was probably an order of magnitude short of a Nobel Prize. My image of success was a full professorship, international recognition for the research I was going to do, and an abundance of graduate students, research grants. (27L)

A development in these scientists' outlooks that becomes more transparent in late-career phases is an intensified rejection of external, science-based definitions of success. Instead, communitarians increasingly turn to themselves, and if to others, then to non-scientists to form estimations of their worth. In effect, they create a community outside of science—which may consist of simply themselves and their spouses, special friends, or family—to help render a kinder and more meaningful judgment in their eyes about "how the career adds up." Thus, in a reversal of scientific norms and how they operate institutionally, non-specialists are called upon to judge the significance of one's work and career and grant an according measure of recognition in the form of esteem, honor, and privilege in this community of non-scientists:

> There's a fellow who's always been a good guy. He's always been a good friend. He is a former department chair [of physics]. Years later, we were talking, and he said, you know, one of the things that he had learned was that quite often people in these kinds of positions will set just extraordinarily high standards for themselves. In some cases, standards that are impossible to reach. And I remember him asking me, "Your friends who are not at the university, or your family members, or whatever, what do they think about this career that you've had up to this point?" And when you think about it, well, you know, my family members think, you know, everything I've done is just fantastic. Other people seem to look at what I've done—I don't know how to describe this—but to see it as a greater accomplishment than the way I have viewed it. And I started to think, maybe sometimes I'm just a little bit too hard on myself. . . . Sometimes we set really high standards for ourselves and maybe we need to look to see, how would your family judge what you've done, or how would non-physicists, non-scientists judge what you've done? I think my definition of success is a little softer. . . . I'm less critical of myself and actually less critical of others. Going back ten years . . . I was looking more for all of us to be perfectionists. . . . I'm not driven anymore just to turn out x number of papers. I really do try to focus on things where to me the problems are very interesting. So you can't turn out publications at the same rate. I also realize I'm not going to work sixty, sixty-five hours a week anymore. I don't feel like I have the energy, but also, I'm not driven in the same way to do that. (34L)

A feature that further distinguishes the ways in which communitarians account for their careers is the low regard in which they hold their institutions. They speak of their universities in adversarial terms; the career comes to consist, institutionally, of a battle waged against ascribed mediocrity that is seen to be produced by an overdeveloped bureaucratic system of academic governance. The one-time scientist, pent up in an iron-like system of comparatively severe constraints, is now nearly seeking release from a cage, as if demonstrating Max Weber's well-known formulation about the pitfalls of advanced bureaucracy.

> INTERVIEWER: What made you decide to retire from physics?
> SCIENTIST: I wanted to get out of teaching. I wanted to teach nothing. I didn't want to do any more teaching.
> INTERVIEWER: Why not?
> SCIENTIST: Tired of it.
> INTERVIEWER: Why?
> SCIENTIST: Thirty-three years. Thirty-three years of teaching. I just got tired of it.
> INTERVIEWER: ... Do you miss being a faculty member?
> SCIENTIST: No. Absolutely not.
> INTERVIEWER: Why?
> SCIENTIST: Because I'm tired of being a faculty member. I wore myself out being a faculty member. I got bored. I got bored being a teacher. I did the same thing over and over and over and over again. At [this] state institution, micromanagement became mandated. Academic freedom was being threatened; it was being restricted. We are losing our academic freedom to teach our courses the way we feel they must be taught in order to produce and generate an educated population of students. It will continue to get worse. We live on a different planet than we lived on twenty years ago, or even ten. We are sacrificing our educational freedoms. To me, it's disgusting. We don't educate anymore. We train. We train students to earn a living. We do not educate students to think any longer.
> INTERVIEWER: ... At what age would you say you were the most satisfied?
> SCIENTIST: Sixty-five. When I left [two months ago], I was most satisfied. I couldn't be more satisfied.
> INTERVIEWER: ... If you were starting all over again, would you go into an academic career?
> SCIENTIST: No, absolutely not. And I wouldn't recommend it to anybody. You're too constrained. We no longer educate, we train. That is the greatest contamination of the intellectual. (27L)

Another of the scientists, retired from a different communitarian institution, said this:

> Since I retired, I'm enjoying not working. . . . I guess if I were starting over again, I probably wouldn't go into physics again. There are more exciting developments in other fields. . . . [The university] was a relaxed environment. There wasn't a lot of pressure to work or to produce, publish, or get grants. Once I reached the full professorship, there wasn't much pressure. . . . I don't think I established myself as an educational researcher to the extent that I would have liked. I think, in part, it would have been helpful to have found some collaborators, and I wasn't able to do that. . . . Probably my biggest complaint was that the administration [of the university] was not supporting the department. When things got better [financially], they didn't restore money to the department that they had taken away. A lot of micromanagement. Toward the end of the career, there seemed to be more paperwork that we had to do related to our job. . . . I felt a bit isolated. . . . I was working in a vacuum. I didn't have colleagues, and I didn't feel that I had the resources to get across the importance that I thought my work had, or could have had, I guess. (37L)

Yet another of the scientists described faculty life in the communitarian world in even less antiseptic terms:

> Our faculty meetings, sometimes boring, but that's typical, consist of mathematical geniuses sitting around a table discussing the problem of toilet paper supply for two hours. (24L)

As these accounts indicate, communitarians in late-career phases exhibit a variety of ancillary career patterns that inform their overall condition. Many say they would not again seek an academic career; others might, but with doubt and reservation. If faced with the prospect of starting an academic career over again, these communitarians frequently assert that they would seek a different type of institution for their careers. But others demurred even at the possibility of what they would do differently in an academic career and instead adhered to a different course of thought, a course for a career away from academia.

> What would have been better for many physicists is if they had been aware, made aware of other opportunities other than just academic positions. That's the problem with most professors. When you have

PhD students, most of the time you're training them to fit into some academic slot unless you consciously make an effort to at least make them aware of other things that they can do. (34L)

Still, with retirement on their minds, they note not having seriously entertained leaving their universities in the past ten years, a finding that again likely speaks more about opportunity costs of leaving rather than a high level of satisfaction that keeps them in place. The calculus is clear in their minds: in their fifties or sixties, what else could they do? Their major calculation, now, is when to retire, helping to further crystallize a utilitarian-type orientation to their work.

But while academic science—or more precisely, college teaching—has remained their calling, it is now a much muted call. Professional aspirations—desires for scientific achievement—are now virtually nonexistent. As earlier passages conveyed, they no longer seek recognition from the scientific community for their work but instead turn to their local institutions, and often even outside of those places, in order find judges who will render, in the words of one of the scientists, a "softer" and kindlier ruling on their lives in science. Moreover, life outside of work is generally seen by them to carry more rewards than life at work.

> I realize I'm slowing down. . . . I don't have a lot of professional aspirations tied to the physics career at this point. I'm thinking about investing time into some other things. . . . I get a little more satisfaction out of spending time with the grandkids [who live nearby]. That, in a way, sort of surprised me a little bit, but you know, it's enjoyable. Fortunately, I've always had a good relationship with my sons, and I find lately I'm spending more time with them. (34L)

Correspondingly, they view the reward system of science skeptically; in large measure, it is, in their eyes and experiences, unfair. In a related pattern, they commonly identify the point of their peak satisfaction as their early careers. Why then? The career patterns suggest three interlocking reasons. First, it was in early-career phases when they contributed their greatest volume of publication productivity. Second, because they were most productive in these phases, they stood the greatest chance of receiving recognition from the scientific community, affirming a sense of honor and positive self-regard. And finally, highly significant career adaptations still awaited them unannounced—they had yet to make the dramatic adjustments in outlook on work that one now sees them making, as they are able to look back on their careers and identify starkly contrasting eras of their lives in science.

One communitarian member of this cohort represents a case that deviates from the typical career patterns. It is a case in which the scientist more closely approximates career patterns characteristic of elites. The scientist may thus be understood as an "elite communitarian." As an elite communitarian, the scientist exhibits a predominant pattern of continuation, the modal means of passage among elites from mid- to late career. At mid-career, the scientist also exhibited elite-like patterns in outlook, orientation, and publication productivity in the career. By late career, he accounted for the evolution in his work in the terms reminiscent of those adopted by elites—those that stress time and progressive research activity; career satisfaction in the present; yet a sense of not enough demand of self and consequent achievement; a desire to do it all over again coupled with the idea that a career in physics, for this type of scientist, will never end.

> I'm working harder than I ever have right now. More hours. Increased commitments. . . . I got back from a research conference last night at eleven thirty and got up here around eight, and I had 150 e-mails. I've got another 250 that are filtered in another pile somewhere, and I haven't gotten through those. . . . I was promoted to [an endowed professorship], so that's about as high as you can go [in institutional rank]. . . . I'm more satisfied with my career than I was ten years ago because of the research.
>
> INTERVIEWER: . . . Would you say you define success differently now than you did ten years ago?
>
> SCIENTIST: I had a goal of getting one hundred publications. I've reached that goal. I have a goal of training this many students. I'm still publishing papers, still doing research. My wife wonders what I'm going to do when I quit. That's probably not going to happen. I don't know what I'm going to do. I'm not even worried about that.
>
> INTERVIEWER: . . . Has there been a point in your career where satisfaction has been the greatest?
>
> SCIENTIST: Right now. It's still very challenging right now. . . . I'm very satisfied with it [my career]. . . . I could have done more. I could have gotten more grants and trained more students or had more postdocs.

INTERVIEWER: If you were starting all over again, and learning what you have about academic careers, would you go into one again?

SCIENTIST: Yes.... It's been very satisfying. I make a good salary. ... You get a lot of freedom to choose what you want to do. That is a big plus.... I could always use more salary, [but] I'm the highest-paid [faculty member] in the department.

INTERVIEWER: ... Do you feel sufficiently recognized for your work?

SCIENTIST: Yes. I'd say so.

INTERVIEWER: ... What would you say are your current aspirations? According to information you gave me ten years ago, you're now approximately sixty-two. You have x number of years to go. What do you want to do?

SCIENTIST: Full steam ahead.... If I'm still enjoying it [in three years, at the age of sixty-five], I'll probably keep doing it.... I think there'll be a time that I'll say good-bye to this university, for the most part. [But] I'll still be doing [physics], I'll [still] be designing things. (25L)

There are no other cases in the sample of communitarian members of this cohort like this scientist, just as there is typically one, or no more than just a few, cases that deviate from predominant patterns that characterize specific cohorts in specific academic worlds.

SUMMARY

By late career, differences among scientists' careers across academic worlds have fully developed. Many of these differences showed their nascent forms much earlier, often in the scientists' first years as university professors. As their careers unfolded, the differences appeared more clearly, such that by mid-career, the point at which the scientists discussed in this chapter were first interviewed, one was able to see clear contrasts. Now the contrasts are even sharper, particularly between scientists in the elite and communitarian worlds. Some of the scientists in the pluralist world, owing to the structure and culture of that world, present less of a contrast and instead resemble elites on one hand and communitarians on the other.

As in the preceding chapter, where the developments in careers by scientists who passed from early to middle phases were dealt, it is possible to codify the major findings among the scientists who have left middle points for the late phases of their careers. Likewise, it is possible to bridge these findings with those generated by the foundational study, while also presenting additional insights, generated by the longitudinal work, about the patterns that careers across academic worlds assume at these junctures in time. Table 25 codifies the principal findings about the scientists' careers as they move from middle to late phases. The table is parallel in structure to that used to codify findings of the youngest cohort discussed in chapter 2.

For elites, the focus of their careers remains research. This characterization, though, carries its own paradox. While the focus of elites' careers is unchanged as such, it is because of this continuity in focus that elites see tremendous productive change in their records of scientific achievement, a finding to be discussed more fully in the final chapter. In short, integration into the institution of science and higher education through research appears to bring about career continuity, both in role performance and in the subjective outlooks scientists form about their careers. Among pluralists, the focus is bimodal: some come to focus on teaching, others continue their focus on research, resembling elites. While communitarians may still publish, their rate of publication productivity plummets, as table 23 demonstrates. By late career, teaching is their dominant role.

While elites remain focused on research, they recalibrate their professional aspirations. Still lofty, their aspirations are not as lofty as they once were, and while these aspirations are perhaps more realizable given individuals' new assessments about their abilities and opportunities to succeed, elites also know that the distance between where they are and their destination is substantial, also an indication of the growth and development that inheres in the act of recalibration. Pluralists' aspirations assume a form that corresponds to the bimodal progression though career phases of this academic world. In some pluralists, professional aspirations subside as they embrace teaching more vigorously. Aspirations remain steady for pluralists whose research remains the focus of their careers. Professional aspirations among communitarians in late career are nonexistent.

In ways that parallel the focus of their careers and the evolving nature of their professional aspirations, elites seek substantial recognition, a career characteristic of possessing a moral orientation to work, which in turn is a function of seeing work as the main focus of life in a constellation of professional and extra-professional roles. Adhering to one of the two predominant modal patterns that mark their world, pluralists may seek average recognition in conjunction with possessing a utilitarian outlook, one that is

TABLE 25. Mid- to late-career patterns of scientists

Career Dimensions	Elites	Pluralists	Communitarians
CAREER FOCUS			
In Mid-Career	Research	Research & Teaching	Teaching
At Late Career	Research	Teaching or Research	Teaching
PROFESSIONAL ASPIRATIONS			
In Mid-Career	Intensify	Diminish	Subside/ Extinguish
At Late Career	Recalibrated	Subside or Remain steady	Nonexistent
RECOGNITION SOUGHT			
In Mid-Career	Great	Average	Minimal
At Late Career	Substantial	Average or Substantial	None
ORIENTATION TO WORK			
In Mid-Career	Moral	Moral or Utilitarian	Utilitarian
At Late Career	Moral	Moral or Utilitarian	Utilitarian
WORK/FAMILY FOCUS			
In Mid-Career	Work	Family & Work	Family & Work
At Late Career	Work	Family or Work	Family & Leisure
ATTRIBUTION OF PLACE			
In Mid-Career	"Haven"	"Happy Medium"	"Stymieing"
At Late Career	"Haven"	"Place to Succumb or Triumph"	"Hapless Bureaucracy"
OVERALL SATISFACTION			
In Mid-Career	High	Medium	Low
At Late Career	High	Medium	Low

(continues)

TABLE 25. (continued)

Career Dimensions	Additional Late Career Patterns		
	Elites	Pluralists	Communitarians
CAREER PROGRESS	As Expected; Better than Expected	Not as Expected; Better than Expected	Not as Expected
WORK INTENSITY	As Hard; Harder	Less Hard; As Hard	Less Hard
OBJECT OF SATISFACTION	Research; Administration	Teaching or Research	—
PEAK SATISFACTION	Present	Early Career; Present	Early Career
REWARD SYSTEM	Fair	Unfair; Fair	Unfair
DEFINITION OF SUCCESS	External	Internal or External	Internal
WORK ATTITUDE	Positive	Ambivalent; Positive	Neutralized
PROMINENT CONCERNS	Funding; Time	Career Direction or Funding	Retirement; Boredom
ACADEMIC CAREER AGAIN	Definitely	Maybe; Yes	No; Maybe
DO DIFFERENTLY	Nothing	More mentoring; Nothing	Different institution
LEAVE UNIVERSITY	No	No	No
RETIRE NOW	No	Possibly; No	Actively Contemplating; Already Retired
OVERALL MODAL PATTERN	Continuation	Regeneration or Continuation	Demise

associated with placing greater stress on family investments of time. On the other hand, pluralists, like elites, may seek substantial recognition, adopt correspondingly similar moral orientations to work. By late career, communitarians no longer seek recognition from external communities but are instead calculating their retirements—an index of their utilitarian orientation to work—and have come to find rewards in leisure and family above those of work.

These patterns lead scientists to view the universities where they work—their attribution of place—in different lights. Elites continue to regard their universities as a "haven," a cultural and structural arrangement of people, beliefs, roles, and resources that facilitates, rather than impedes, their work. For pluralists, the university becomes a "place of routines" in which one can either succumb or triumph. Those who seek triumph nevertheless are aware of obstacles they perceive their institutions creating for their work, as indicated by the scientists who believe they must always work to overcome "being held back." In their minds, the institution, and the routines found within it, are to be overcome through stamina and strategic individual maneuvering. Communitarians view their institutions as "hapless bureaucracies," designed more to impede than to facilitate work.

Along these lines, overall satisfaction among elites may be said to remain high, in comparative terms. Likewise, for pluralists it may be said to be medium, even for those whose modal career patterns more closely resemble elites; in their estimations, their institutions create "a lot to put up with," which compromises a fuller realization of satisfaction. They also believe their institutions hinder the reach of their recognition and external rewards. For communitarians, overall satisfaction is comparatively low. Many of them find new rewards in teaching and faculty-student interaction, but these prove not enough to overwhelm all that they see has been lost.

Additional dimensions of scientists' careers, evinced through the follow-up work and identified in the second part of table 25, help to place career patterns in further context of their respective academic worlds. Elites find that their careers have progressed as expected or better than expected. In their work intensity, they claim to work as hard or harder than they did ten years ago. They derive greatest satisfaction from their research and, now increasingly in late-career phases, from administrative roles. In bimodal fashion, pluralists regard their careers as either not having gone as expected or better than expected. They see themselves, bimodally, working less hard or as hard. Teaching or research, a clearer division from ten years ago, is the principal object of their work satisfaction. For communitarians, the career has not gone as expected, and consequently, they see themselves working less hard. There is now no clear object of satisfaction at work.

Elites in late career claim to have reached their peak satisfaction; they see themselves living it presently. For pluralists, peak satisfaction is found in the present or the early career, and for communitarians unmistakably the early career, when they were most productive in research, most rewarded, and least blunted in their aspirations.

These sentiments are associated with different estimations of the scientific reward system. Elites view it as fair, not flaw-free, but, on balance, a just barometer of scientific achievement. Pluralists view it either on these terms or as unfair, the manner in which the comparatively least recognized communitarians regard the system. In this vein, attitudes about the reward system appear in a positive correlation with the benefits that scientists have derived from it.

Overall, elites' attitude toward work is positive, despite their most prominent concerns about funding to initiate and execute their work on the one hand and the amount of time to complete their work on the other. Pluralists either share these views or have adopted an ambivalent attitude toward work enjoined by an overarching concern with the direction in which their careers may, or may not, go. Attitudes toward work among communitarians have neutralized; they are putting in time toward retirement, and in the process, often attempting to overcome boredom.

Unhesitatingly, elites would seek an academic career again. Some pluralists, those concerned about the state and direction of their careers, give more pause. Communitarians would do something else entirely, or give an academic career the barest reconsideration. And if they did it again, nothing would be done differently, according to elites, as with pluralists whose modal patterns resemble elites. Other pluralists would seek more mentoring from senior colleagues to help ensure greater research continuity, which they sense to be a more reliable protection from the drift that now plagues them. Of those communitarians who might consider an academic career again, they would seek it in a different kind of institution, one where research was more ably supported.

Despite career concerns and criticisms about their institutions, none of these scientists have wanted to leave their universities. As explained earlier, this is a finding that likely informs the level of sunk costs scientists have made toward their work by this point in time, or because of the real satisfaction they find in their work. Elites would not retire now if given the chance;

some pluralists would; communitarians are actively contemplating it, and the ones who have already retired are found in their midst.

Overall, elites observe a modal career pattern of continuation in aging from mid- to late career. In these phases, pluralists are bimodal, characterized either by continuation like elites or by regeneration, the process by which scientists seek to recover from comparatively slow periods and in which they seek greater clarity and direction in their careers. Communitarians are marked by demise. Short of retirement, they are in the final phases of completing, and in coming to terms with, the realities of their active careers.

In turn, the conclusion of careers is the phase in this work that has been reached. Scientists have been followed into their late careers and have spoken about how they interpret this set of passages that brings them toward a major marker, and a potentially major set of turning points to further characterize, and differentiate, careers in science. To complete the view of careers, as scientists see them, the discussion turns to those who have completed this final passage.

CHAPTER FOUR
Late- to Post-Career Passages

The third and final cohort of the study has proceeded from late to later phases of their careers. "Later career" carries its host of multiple meanings. It includes people who are well-advanced in their careers and those who have retired but have done so in varieties of ways. Retirement from science, as will become clear, means altogether different things, and entails substantially different ways of life, among scientists.[1] To be inclusive of scientists and the patterns of careers they exhibit at these points in time, one may refer to the passages they make from late to *post*-career, the latter situated as the outermost temporal point where a divide demarcates scientists' occupational careers from their gainful lives away from them. As in the consideration of the two younger cohorts, the discussion will center on how scientists pass from late- to post-career phases in various ways, underscoring cohort differences in occupational aging. And it will also be possible to see further how scientists vary at the end of their careers among the three types of academic worlds—elite, pluralist, and communitarian.

The order of discussion will follow that of the two previous chapters. I will present a professional profile of the eldest cohort to place it in context and to help situate its members' accounts of their careers. I will then summarize these scientists' career patterns from late career, as found in the foundational study, to establish a basis on which to draw further comparison and to build knowledge about the unfolding of scientific careers developed from the longitudinal work. Finally, I will turn to discussing patterns that scientists exhibit as they pass from late to post-career across the three worlds of science.

PROFESSIONAL PROFILE

In table 26, I provide a summary of descriptive professional information on members of the late- to post-career cohort of scientists, in order to place

TABLE 26. Cohort characteristics: the late- to post-career cohort[a]

Characteristics	Elites	Pluralists	Communi-tarians	Overall Average
AGE & EXPERIENCE[b]				
Age Range @ 1st Int.	59–67	52–67	53–63	55.0–66.0
Avg. Age @ 1st Int.	64.2	59.0	60.1	61.4
Age Range @ 10Yr. Fw-Up	69–77	62–72	63–73	65.0–76.0
Avg. Age @ 10Yr. Fw-Up	74.2	67.0	70.1	71.1
Range Yrs. Exp. @ 1st Int.	21–38	21–38	27–34	23.0–36.0
Avg. No. Yrs. Exp. @ 1st Int.	31.0	30.0	30.2	30.2
Range Yrs. Exp. @ 10Yr. Fw-Up	31–48	31–41	37–44	33.0–46.0
Avg. No. Yrs. Exp. @ 10Yr. Fw-Up	41.0	35.0	40.2	37.0
JOURNAL PUBLICATIONS: QUANTITY[c,d]				
Maximum @ 1st Int.	272	89	100	153.7
Minimum @ 1st Int.	44	19	9	24.0
Mean @ 1st Int.	111.0	60.7	39.1	69.7
Maximum @ 10Yr. Fw-Up	536	116	111	254.3
Minimum @ 10Yr. Fw-Up	44	19	9	24.0
Mean @ 10Yr. Fw-Up	159.1	82.4	52.0	98.0
JOURNAL PUBLICATIONS: QUALITY[e,f]				
≥80% published work/1st Int.	67.0	67.0	43.0	59.0
≥70% published work/1st Int.	78.0	83.0	43.0	68.0
≥60% published work/1st Int.	100.0	83.0	71.4	84.8
≥80% published work/10YrFwUp	56.0	80.0	40.0	59.0
≥70% published work/10YrFwUp	67.0	80.0	40.0	62.3
≥60% published work/10YrFwUp	78.0	80.0	40.0	66.0
ROLE PERFORMANCE				
Avg. No. Papers @ 1st Job[g]	4.5	5.0	2.1	4.0
Avg. No. Papers @ Tenure[h]	11.0	14.3	8.3	11.2
Avg. No. Paper @ Full Prof.[i]	22.0	24.3	17.0	21.1
Avg. Time to Tenure (in years)[j]	5.0	4.0	5.3	4.8
Avg. Time to Full Prof. (in years)[k]	5.0	5.3	5.6	5.3

Notes:
[a] Source: Each scientist's curriculum vitae.
[b] Elites: N = 9; Pluralists: N = 6; Communitarians: N = 7. Includes all members of the foundational study sample. Data for longitudinal items, pluralists, exclude one deceased member

(continues)

TABLE 23. (continued)

of the sample. Range and average years of experience timed up to year of professors' formal retirements.

[c] The number of publications for each scientist includes all published journal articles, scientific journals being the primary medium through which scientists disseminate their research. The number excludes the following: books; textbooks; book chapters; edited volumes; conference proceedings; invited and contributed papers; book reviews; encyclopedia, world book, and yearbook entries; and articles listed on the individual's curriculum vitae as "submitted," "in press," "accepted for publication," "in preparation," and so on. If the same journal articles was published multiple times (in different venues), it is counted only once.

[d] For foundational study: Elites: $N = 9$; Pluralists: $N = 6$; Communitarians $N = 7$. For longitudinal study: Elites: $N = 9$; Pluralists: $N = 5$; Communitarians: $N = 5$.

[e] Percentages of scientists' publications appearing in "major physics journals," those journals to which the physics community assigns greatest value. These include, in alphabetical order, Astronomical Journal, Astrophysical Journal (including Supplement Series), Astrophysical Letters and Communications, Europhysics Letters, Geophysical Research Letters, Icarus, International Journal of Modern Physics (including series A, B, C, D, E), Journal de Physique (including series I, II, III, but excluding IV), Journal de Physique Lettres (now incorporated with Europhysics Letters), Journal of Chemical Physics, Journal of Geophysical Research (including series A, B, C, D, E), Journal of Mathematical Physics, Journal of Physics (including series A, B, C, D), Lettre al Nuovo Cimento (now incorporated with Europhysics Letters), Nature, Nuclear Physics (including series A, B), Physical Review (including series A, B, C, D, E, L), Physics Letters (including series A, B), Physics of Fluids, Review of Modern Physics, Science, and Solid State Communications.

[f] For foundational study: Elites: $N = 9$; Pluralists: $N = 6$; Communitarians $N = 7$. For longitudinal study: Elites: $N = 9$; Pluralists: $N = 5$; Communitarians: $N = 5$.

[g] Average number of journal articles published by the time scientists were appointed to their first job. Other publications excluded. "First job" is defined as appointment as assistant professor or visiting assistant professor. Elites: $N = 6$; Pluralists: $N = 6$; Communitarians: $N = 7$. Cases do not agree with total cohort numbers because of unavailable data. Scientists who began their careers outside of academic science (e.g., as industrial scientists) are excluded.

[h] Average number of journal articles published by the time scientists earned tenure. Other publications excluded. Elites: $N = 6$; Pluralists: $N = 6$; Communitarians: $N = 6$. Cases do not agree with total cohort numbers because of unavailable data. Scientists who began their careers outside of academic science (e.g., as industrial scientists) are excluded.

[i] Average number of journal articles published by the time scientists were promoted to full professor. Other publications excluded. Elites: $N=6$; Pluralists: $N=6$; Communitarians: $N=6$. Cases do not agree with total cohort numbers because of unavailable data or because scientists are not full professors. Scientists who began their careers outside of academic science (e.g., as industrial scientists) are excluded.

[j] Average time in years it took scientists to receive tenure. Elites: $N = 6$; Pluralists: $N = 6$; Communitarians: $N = 6$. Cases do not agree with total cohort numbers because of unavailable data. Scientists who began their careers outside of academic science (e.g., as industrial scientists) are excluded.

[k] Average time in years it took scientists to earn promotion to full professor. Elites: $N = 6$; Pluralists: $N = 6$; Communitarians: $N = 5$. Cases do not agree with total cohort numbers because of unavailable data or because scientists are not full professors. Scientists who began their careers outside of academic science (e.g., as industrial scientists) are excluded.

these scientists and their careers in greater context. The table is parallel in structure to the professional data on the two younger cohorts presented in tables 20 and 23.

The scientists, grouped as late to post-career, range in age from 65 to 76 years; on average they are 71.1 years of age. Members of the elite subsample are somewhat older on average (74.2 years) and pluralists, on average, somewhat younger (67 years).

The scope of experience of the scientists, coupled with time, conveys further the temporal location of phases in which they now find themselves. They have accumulated an overall average of between thirty-three and forty-six years as university professors. The average is thirty-seven years. Further, as information in table 8 conveyed, these scientists, with limited exceptions, have worked over this period of time at the same institutions, thus providing a relatively stable set of organizational contexts in which to examine career development and the shaping of identity and personal-professional perspectives over time.

As the scientists have advanced in age and proceeded to new phases of their careers, so, too, have their records of publication. Elites of this cohort have published an average of 159.1 articles, pluralists 82.4, and communitarians 52. While these records are substantial in absolute terms, comparison with the other cohorts is instructive. Elites of the eldest cohort outpublished their peers in the mid- to late-career cohort by an average of only 11.1 papers. Pluralists of the eldest cohort actually published an average of 45.6 fewer papers than their counterparts in the mid- to late-career cohort, a difference that is attributable to an especially prolific mid- to late-career pluralist and to an overall greater push to publish among younger cohorts of scientists. Communitarians of the eldest cohort published an average of 16.8 more papers compared with their counterparts in the mid- to late-career cohort. Indeed, with their average publication productivity of 52 papers, communitarians of the eldest cohort published more closely on par with communitarians of the youngest cohort (an average of 50 papers). The evidence is suggestive of the historical conditions under which academic careers have transpired. The pressure to publish has been felt most keenly by members of the youngest two cohorts, who entered science after 1970. Members of the eldest cohort, who entered science prior to 1970, publish substantially (especially among elites and pluralists), but have done so at a markedly lower rate compared to younger scientists. That several of these eldest scientists, particularly from the communitarian world, began their careers with a greater teaching focus compared with their contemporaries further explains why one observes a relative muting of their publication productivity.

The placement of published work into what the science community deems the leading journals also has fluctuated among members of the late- to post-career cohort. Elites who published 80 percent or more of their work in the leading journals declined from 67 percent to 56 percent since late career; for pluralists, it rose from 67 percent to 80 percent; and for communitarians, it declined slightly from 43 percent to 40 percent. Using the quality threshold of 60 percent, the quality of publications declines overall among members of the cohort. At the time of the first interview, when scientists were in late-career phases, 84.8 percent of them published at least 60 percent of their work in leading outlets. Ten years later, when scientists reached the later-most phases of their careers, 66 percent of them published 60 percent or more of their work in the leading journals. This constitutes a somewhat lower average compared with scientists in mid- to late-career phases (72 percent) and a higher average compared with scientist in early- to mid-career phases (56.4 percent), suggesting that the youngest scientists feel greatest pressure to publish and to do so sometimes at the expense of quality.

Differences in the pressure to perform in research and publication is nowhere more evident than in the data on role performance in the last panel of table 26, and in the last panel of tables 20 and 23 for the two younger cohorts. Scientists in the late- to post-career cohort published an overall average of 4 papers prior to obtaining their first academic positions. For members of the mid- to late-career cohort, this average was 11.1; for the youngest cohort, 14.3. To earn tenure, members of the eldest cohort published an overall average of 11.2 papers. Members of the mid- to late-career cohorts had published an overall average of 23 papers to obtain the same outcome. Scientists in the early- to mid-career cohort published an overall average of 32 papers prior to tenure, a marked generational difference between themselves and their eldest peers, and even a substantial difference between themselves and their mid- to late-career peers. The data make unambiguously clear how high, in a span of approximately thirty years, the bar has been raised in order to secure one's position in academia.

Security in the form of academic tenure is one threshold, advancement in the form of promotion to full professor is another. Members of the eldest cohort had published an overall average of 21.1 papers prior to promotion to full professor. By contrast, members of the mid- to late-career cohort had published an overall average of 41.5 papers, and members of the youngest cohort had published an overall average of 44 papers to achieve the same outcome. Here, too, generational differences in performance expectations are transparent.

The findings are made even more remarkable when viewed in the time in which scientists made their performances. Members of the eldest cohort were promoted to tenure after an overall average of 4.8 years. Members of the mid- to late-career cohort took an overall average of 4.4 years to reach tenure. Members of the early- to mid-career cohort earned tenure after an overall average of 5.8 years. Thus, while members of the youngest cohort published an overall average of 20.8 more papers for tenure compared with their most senior counterparts, they did so with an overall average of but one additional year as university faculty members.

As the foundational study documented, 94.5 percent of the youngest cohort held a postdoctoral appointment, compared with 68.8 percent among members of the middle cohort, and 54.5 percent among members of the eldest cohort (Hermanowicz 1998, 148, table 11). Fully 63.6 percent of the youngest cohort held more than one postdoctoral appointment, compared to 3.8 percent and 1 percent among members of the middle and eldest cohorts, respectively (Hermanowicz 1998, 148, table 11). Thus the postdoctoral phase of a scientific career, now nearly universal in the physical sciences and increasingly common in the biological sciences and in some (especially quantitative) areas of the social sciences, such as demography, establishes a period before onset of a tenure-track position to establish publication records prior to evaluation for tenure.

The thrust of patterns is similar for time to advancement to full professor among the cohorts. The eldest scientists were promoted to full professor after an overall average of 5.3 years, the youngest scientists also after 5.3, and the middle cohort after 5.8 years. Thus the youngest scientists published an overall average of 22.9 more papers compared to their most senior counterparts and approximately the same number of papers compared to the middle cohort in order to rise to full professor, again suggestive of a sharp generational divide between pre- and post-1970 onset of careers in higher education.

LATE-CAREER PATTERNS

At the time of the foundational study, when scientists in the present cohort were in late phases of their careers, differences in outlook and orientation

TABLE 27. Late-career patterns of scientists

Career Dimensions	Elites	Pluralists	Communitarians
CAREER FOCUS			
In Late Career	Research	Research & Teaching/ Teaching	Teaching
PROFESSIONAL ASPIRATIONS			
In Late Career	Remain Steady	Subside	Not present
RECOGNITION SOUGHT			
In Late Career	Great	Average	None
ORIENTATION TO WORK			
In Late Career	Moral	Moral or Utilitarian	Utilitarian
WORK/FAMILY FOCUS			
In Late Career	Work	Family & Work	Family & Leisure
ATTRIBUTION OF PLACE			
In Late Career	"Haven"	"Happy Medium"	"Stymieing"
OVERALL SATISFACTION			
In Late Career	High	Medium	Low

to their work presented numerous contrasts. By that time, they had worked as university professors an average of between twenty-three and thirty-six years. They ranged in age from fifty-five to sixty-six. In short, experience had expressed itself in systematically distinct ways across the organizational settings in which the scientists had pursued their careers. Table 27 presents key generalizations of late-career patterns for the scientists across their academic worlds.

In late career, elites remained focused on research—this component of their professional roles remained most prominent and most pressing. Their aspirations remained steady, and they continued to seek great recognition for significant scientific achievements. The magnitude of recognition they sought appeared qualitatively greater than that sought by their younger counterparts, who subsequently advanced into late phases. This younger elite cohort sought substantial recognition but recalibrated their aspirations by late career to become more aligned with what they thought was more realistic to achieve. Members of the present cohort in late career were less compromising. It is likely that the auspicious times under which their careers had developed, particularly the comparative ease with which they had been able to secure funding over the majority of their careers, helped to foster and sustain heightened expectations for achievement.

Not surprisingly, their orientation to work remained moral: science remained its own end. And even when much had already been achieved in the professional arena, work—among a variety of other life arenas in which time and identity could be invested—remained their central focus. They continued to speak of the institutions in which they worked as "havens," places that they believed and understood to facilitate their work and progress in research. In comparative terms, their overall satisfaction could best be characterized as "high."

Among pluralists, the focus of the late career was a combination of research and teaching or exclusively teaching, reflecting a bifurcation often found in this academic world. In general, aspirations for members of this cohort subsided, and they in turn sought what may be characterized as average levels of recognition, an apparent corollary of their shift in focus away from research toward teaching.

Where research was a meaningful component of the career, scientists kept a moral orientation to work. Where research dissipated or disappeared as a meaningful component of the career, scientists typically adopted a more utilitarian outlook on their work. The career became more a means to a set of financial ends, a way to make a living. Family more readily entered pluralists' accounts of their occupational careers, another apparent corollary of the weakening hold of work on scientists' identities. For these reasons, pluralists understood their institution as a "happy medium," and this corresponded to the level that could best characterize their overall satisfaction.

Their university was not, in late career, a place to "succumb or triumph," as it was for their younger counterparts who had advanced to late career. It appears as though their younger counterparts made greater demands on themselves for scientific achievement, likely because their institutions were making greater demands on them. The attitudinal result, for pluralist members of the mid- to late-career cohort, was a greater expectation for their institution to help them satisfy these achievement demands. Such demands appeared less intense among members of the present cohort, who entered and were conditioned by their pluralist institution under more moderate performance expectations. Nevertheless, the existence of (comparatively modest) research expectations in an environment in which members could not fully realize them produced an overall level of satisfaction that could best be described as "medium."

For communitarians, teaching was their principal focus at late career. Professional aspirations tied to research were no longer present: their institutions could not sustain them. Consequently, communitarians no longer sought professional recognition. If research was done, it was research that did not require funding, was out of a professor's general interest, was viewed as a sporadic, side activity, and, further still, was one not necessarily seen to completion or publication.

Academia was a job. Communitarians embraced a utilitarian outlook on their work. To the extent they may have identified themselves as scientists or physicists, it was part of their biography, a past identity that served as a point of contrast from the present. Their focus among a possible constellation of roles was family and leisure; work was no longer prominently in the picture. They regarded their institutions more as "stymieing" than a "hapless bureaucracy." The distinction centers on a proactiveness: those who work among hapless bureaucrats still navigate the bureaucracy. Those who simply feel stymied have less hope and are "putting in time." At late career, communitarians' overall satisfaction could best be characterized comparatively as "low."

Ten years later, outlooks and orientations have evolved. So, too, has professional status. Of the nineteen interviewed members of this cohort, twelve have officially retired. Retirement status is distributed among the worlds. Six of nine elites, three of five pluralists, and three of five communitarians have retired. The present cohort allows for an especially privileged position: to hear from scientists their perspectives on careers that are now complete. Yet, as will soon be discovered, that final word—*completion*—will prove an elusive reality to scientists most integrated in the institutional goals of science, even at the end of their careers. I turn to what retirement, and a host of other career experiences, mean to scientists having passed from late to as far as post-career.

ELITES

In anticipating their future careers and what retirement might bring, members of younger cohorts, particularly elites, typically expressed the view that they would remain active scientists, that retirement would free them of teaching duties and, if anything, allow more time for scientific work. Such perspectives—like all perspectives formulated within the structure and culture of a social system—are not random, but are instead products of anticipatory socialization: scientists look at their older peers in order to ascertain how careers are to unfold within the cultural confines of their specific academic world. Thus, among its possible meanings, retirement among elites customarily connotes a liberated phase of work, in which scientists actively publish, travel professionally, serve on national and international panels, advise and sometimes sponsor graduate students and postdoctoral researchers, and, perhaps rarely, teach (and then usually at the advanced graduate level). Is this an illusion or is it true? Is such anticipatory socialization created out of thin air, a fantasy for how the career can end, or does it have an empirical basis?

The data provided by members of the eldest cohort convey quite clearly that this is not only how they envisioned retirement but also how they in fact spend it. Those who make a "clean break" from work are a rare exception in this world of science. When talking about or describing life in retirement, not a single member of this world remarked along the line, "When I retire, I'm going to leave science and never look back." As it turns out, most scientists in this world cannot leave science because it is what defines them: in this world, there is no "in" or "out" of science. Thus one finds a ubiquity about the following scientist's account:

> INTERVIEWER: I'd like to see how your perspective on work has changed in the last ten years.
>
> SCIENTIST: I would say it hasn't changed at all. I enjoy it, in spite of retirement in '96 [two years after the first interview, eight years prior to the second interview].
>
> INTERVIEWER: How is it different now from before you retired?
>
> SCIENTIST: I don't teach, but that's about all, and I also don't have any obligations to serve on committees or what have you. I come in every day and work on my research.... I don't have any students. It costs about forty thousand dollars a year to support a graduate student; that

takes quite a lot of proposal-writing and getting contracts to pay for the graduate students. I don't have that burden anymore.

INTERVIEWER: Has your outlook on your career and what your career means evolved since retirement?

SCIENTIST: No, I wouldn't say so. I just plug along at the research that I've enjoyed over the years. The research directions have changed to some extent, but it's fun. And when I finish something, I publish it.

INTERVIEWER: . . . How would you fill in the sentence, "I am more X and less Y compared to ten years ago," before you retired?

SCIENTIST: That's rather abstract. Well, I'm more engaged in research and less in teaching, but that's kind of simplistic. I wouldn't say my life has changed very much, except for the shedding of responsibilities and more just pure pleasure in doing my research. (13L)

Or, as another of the scientists commented:

I'm still trying to accomplish things. You know, when you reach [the age of] seventy-three, one thing you think of—if you're sensible—is how you're going to wind down. And I have no simple answer to that. I'll tell you a story which I think is pertinent. When I was a new assistant professor, some student asked me, "What's life?" I don't know how this came up. Without hesitation, I said, "Activity." That was a very interesting answer for a guy in his early thirties, isn't it? The more I think about that, and I remember it every once in a while, the more I marvel because it just came out of me. You like to get important things done, and one takes a lot of pride out of accomplishments. . . . I retired at age seventy. I forgot how many years into it I am [three]. But that just means I retired from formal classroom teaching. (1L)

A female scientist, age seventy-four, not retired, who is among the most productive members of the sample of scientists, described the rhythm of her career. She is the fourth of the four women covered in the study, and the only woman in the study who is a member of the eldest cohort of scientists. In ways consistent with the career patterns of the other women scientists in the study, her career patterns parallel not only those of the elite academic world of which she is a part, but the most successful among that subset. A frequent traveler on behalf of science, her schedule had little room to overlap with that of an interviewer of scientists. I managed to make our schedules overlap only by extending my visit to the university an additional weekend, to Monday, by which time she would have returned from Japan. On that Monday, the only time to meet was at 7:00 a.m., prior to meetings with students that would consume her morning, and before she headed out-of-town again the next day:

> I come in usually around 6:00 a.m., 6:30 a.m., and leave about 5:30 p.m., 5:15 p.m. I'm here [at the university] about half the time [of the year]. December was a light travel month because of the holidays. I only went to one foreign country—Sweden. In January, I had a really big load: Taiwan, UK, and Japan, in that order. It would have been nice to have it more continuous. I was supposed to go to Chile but couldn't fit it in, so it was only a conference call. [Looking ahead], Utah is the first week of the month, and from Utah I go to the West Coast—I have a panel review Academy meeting. I have to do some homework for that, get organized. I leave tomorrow. From there, I'm supposed to go to Brazil. When I come back, I have to give a plenary talk at a conference in Florida. Right after that, I go to Arizona. I come back here for three days or two days. I make many trips to New York. . . . If I never wrote another paper, it wouldn't be so bad. But I know I'm going to write many more, because I have many in the pipeline, things that I'm working on. It's hard to imagine a time I won't be doing this. (14L)

Finally, a scientist who placed this orientation to work matter-of-factly:

> INTERVIEWER: How do you think your perspective on work has changed over the past ten years?
> SCIENTIST: Not a bit.
> INTERVIEWER: Not one bit?
> SCIENTIST: Not one bit. I'm now retired. I'm emeritus.
> INTERVIEWER: Wouldn't that create some kind of change in your perspective?
> SCIENTIST: I'm not teaching anymore. I still have two graduate students. I think there's no change in my perspective whatsoever. (10L)

At the same time, the eldest elites tend to remark that they work less hard or about as hard as they did ten years ago and that, overall, they are now—for the first time in their careers—less ambitious compared with prior points in time. These characteristics, combined with others to be described below, contribute to a predominant theme of *attenuation*: scientists remain active in science, but less intensely. It is the sense that they continue to do science out of habit, as opposed to the realization of specific goals.

> [I work] less hard because I don't have to prepare lectures. I don't have to deal with committees and so on. I would say I'm working less hard than I did when I was a regular faculty member. . . . [And] I would say [I'm] less ambitious. When I was a regular faculty member, I worked very hard to obtain major projects of which I was the principal investigator. That's a very competitive field, so I worked very hard at that sort of thing. Now I don't do that, so I would say I'm less ambitious. I don't strive to take charge of the major programs. (13L)

Another scientist, age sixty-nine, not retired, repeated these common findings:

> INTERVIEWER: Do you sense that your outlook on your career has changed?
> SCIENTIST: Not much, really. I was surprised that the choice of physics that I made would remain good. It was better than I thought.
> INTERVIEWER: Have you become more active, less active, or are you as active as you were ten years ago?
> SCIENTIST: That's a good question. Of course, age takes a toll. But I'm just getting a paper out with my name on it. It was all done by myself. Medicine has changed our lives. My sight operations—I see better than ten years ago. Certainly I am slower, definitely slower. But in terms of being able to do original work, I have not found that has decreased
> INTERVIEWER: . . . Have you ever thought of just stopping?
> SCIENTIST: Not really. A close friend of mine, [an academic,] died a year and a half ago. He had a bad form of cancer. A few hours before he died, he woke up and was lucid, and he began to work. I think I can go on, as long as I can do something original. It's important to do something which was not done before. That's important.
> INTERVIEWER: So you see yourself as never stopping?
> SCIENTIST: Well, as long as your mind can assist you. I really think the secret is to keep active in your research—not retire from it. (8L)

While, in the accounts above, elites' focus on research remains transparent and thus consistent with earlier phases of their careers, the latter-most career phases bring about a qualitative adjustment in their orientation to work. Where once morally oriented to work, in which external definitions of success motivated a search for great levels of recognition, now elites may be better described as intellectually oriented to their work. This does not mean that they no longer view science as its own end. To the contrary, this view is deepened, but they are now staked less competitively in a search for recognition, even though they continue to uphold external definitions of success. Science remains a serious enterprise, but elites now temper their professional aspirations for even greater achievement. An additional hallmark of this shift is a somewhat greater embrace, or at least a greater discussion, though mild, reserved, and nearly reluctant, of leisure.

> I do a lot of writing. Since I last saw you, I've written about two hundred thousand words about English history. It may or may not be published. I think it's easier than astrophysics, but that was sort of to fill in the time after I stopped teaching. (10L)

> I'm cooking more. I spend more time with my wife now, and I'm exploring. If I was forced to walk out of here and not see the office again, like what happens in companies, that could be awful. (3L)

The great degree to which elites are integrated in the institution of science is conveyed by these varieties of career characteristics: an apparently indefatigable commitment to and engagement in research, the continual adoption of external definitions of success, and an orientation to work that, if not necessarily moral, is intellectual and, above all, scholarly and consistent with their profession's institutional goals.

By the latter-most phases of elites' careers, however, the great degree of this integration in science also produces what might be seen as an unexpected ambivalence and dissatisfaction with their career. Many of them experience a reversal. Having been so closely committed to scientific research, having so vigorously pursued professional goals and achievements, having so faithfully, on balance, believed in the institutional operation and allocation of rewards and recognition, many elites now see themselves as having fallen short, of having not achieved enough in their own minds. Sensing an end, they have come to realize, for the first time in their decades in science, that a new level of achievement will elude them. A reversal of positive sentiment toward work and career, so characteristic of early-career communitarians, is new to elites at late career, and one is able to see striking changes in outlook. From a scientist, age seventy-three, retired, whose career spanned forty years at a renowned American research university:

[The transition to retirement] was a little frightening. I did it voluntarily, and it is basically okay. It basically went the way I thought, but it wasn't quite as comfortable as I thought it would be. It's like being a lame duck, to attract students and things like that, finally it just becomes impossible. You lose your voice in departmental affairs. I still have my office, which is nice. At first, the Department of Energy wanted to take my research funding away. I had to shout at them on that. That was unpleasant. Immediately, the program officer said, "Oh, [Smith] is retiring. We can give you less money." That was bad and made worse by the fact that it took a while [to resolve]. That wasn't a lot of fun. . . . [I haven't made] some great discovery. Now I feel it would have been nice if I tried a little harder, to take a problem I can solve, just take a chance and try. . . . Sooner or later, you've got to really retire. I've had to struggle with funding and being a lame duck. But it's inevitable. Whether you're officially retired, people know you will sooner or later. They know your age. (1L)

From a scientist, age seventy-five, retired, whose career similarly spanned four decades at another prestigious research university with an illustrious history of research:

Not many people make great contributions in science and physics. Mostly, they're incremental contributions, and one doesn't always see a tremendous value in that. If you find some little thing that's new, it is new, but it's not advancing the forefront of understanding and knowledge. There's a lot of work that's done because people want to work, it's like a hobby. Maybe there is some self-delusion in feeling that you're being a significant contributor to science. It's just [pause] you have been trained, you know this field, when you're an expert in something, you tend to take pride in it, and you tend to continue doing it. But I don't think it's always very significant in the grand scheme of things. . . . I could have worked harder to become a better professional physicist. . . . At some stages in my career, I could have easily done better. It would have made a difference. It might well have been a significant difference. . . . If I had worked harder, it would have given me a little more status. I would have accomplished more in the field. . . . I'm not striving [now] for any particular goal. I'm trying to live an active life intellectually. I've got a new grandson. I'm interested in what my sons do. I'm hoping they do well. I'm hoping I can be healthy enough so that I can see them succeed. What further goals I have, just to keep up with things. . . . But I don't have a grand goal at this stage of life. I want to live a stimulating and comfortable life. And maybe I can contribute a little bit to what's around me. (21L)

From a Nobel Prize winner:

Everything I do, I don't feel that I'm as successful as I should be. I'm never really satisfied with anything I do. This is a general feeling on my part. Even when I do research, even when I do research that has been well-recognized. I always think, I could have done it better had I done this or that. I have the same feeling when I prepare talks. I generally feel as though I'm trying to push a rock uphill. It's very difficult. I never quite push it far enough. . . . I feel very fortunate to have been in the right place at the right time to do something significant—that makes me feel pretty good from that point of view. It's just that, I could always do things better, and I'm never really satisfied with what I do. I don't know whether you've ever felt that, but that's the way I feel very often. (16L)

From a renowned astrophysicist, retired, age seventy-four:

I'm conscious of that fact that in the science world, there are standard success [markers]. I have a bunch of colleagues who belong to the National Academy of Sciences. I'm not in that league. I have other colleagues in the American Academy. I'm not in that league. . . . I was never asked, partly because I think I didn't grab a specific aspect of astrophysics and push it hard. When people looked at actual science accomplishments, they weren't there. . . . It bothers me. I look back and say, I could have done things differently. On the other hand, I don't think I'm as smart as some of those people. . . . And there are times when I look back where I didn't—I mean, in the '60s, there were huge discoveries to be made. They were at the fingertips of all of us. Discovery of neutron stars, pulsars. And a few people ended up discovering them more or less accidentally. I'm not alone, I'm sure, saying if I had only changed the course five degrees and looked into this instead of that. . . . You can see life moving on. I'm conscious of it. I don't want to live forever. I get tired of even thinking about living a life again. I just get tired thinking about it. I was biking in from [one of the suburbs] on my bicycle six miles to my office and back every day, dealing with these rocket programs and crises and whether they're going to launch or not launch. And the funding. Thank God I don't have to do that again. (3L)

In these discussions, scientists were serious, pensive, reflective. They conveyed a sense of regret and melancholy, a sadness that the career, and indeed life, was coming to an end, and a painfulness in what was not achieved. In general, the prominent concern for elites in these phases of their careers was *professional standing*—where they saw themselves having ended up. Elites had always been willing to discuss professional standing at earlier career phases. Now, in an associated finding, it was most difficult for elites, those for whom professional standing mattered most, to discuss. In isolated cases, the discussion prompted high tension and an unanticipated near collapse of the interview. An internationally renowned scientist, age seventy-six, formally retired:

INTERVIEWER: Do you have a fear of not being able to do science?
SCIENTIST: No, not a bit. As you get older, you're not as smart. You have to face that. As long as you can keep up, that's fine. It has to do with keeping track of the major scientific questions, and you'll know it when you can't do it anymore. I hope I'll know it, but that point hasn't been reached yet.
INTERVIEWER: What's the indication that people will know?
SCIENTIST: You know it yourself, I hope, first. There are some cases where others notice it. It's when you clearly don't understand. I know when I don't understand something, and if I don't understand it, either I can understand the explanation, or I probe into it. If I don't understand even the probed explanations, then that'll be a bad sign. I have a personal criterion. I'm a member of the National Academy of Sciences. We get an assessment list to vote on possible members, and if I reach the point where half the names are unfamiliar to me, it's time to get out. Very brutal, straightforward criteria. There's no sense of voting on a list where you don't know the people. You're not going to know everybody in the astronomy division, but you better know most of them.

INTERVIEWER: ... Are there ways in which you feel you haven't been successful?

SCIENTIST: No, no. You win some, you lose some. Everybody knows that.

INTERVIEWER: ... Is there anything you wish you had achieved that you haven't?

SCIENTIST: Oh, lots of things. But I'm not going to tell you about them.

INTERVIEWER: Why not?

SCIENTIST: It's an absolute statement. I'm not going to tell you about them.

INTERVIEWER: Because they're not important?

SCIENTIST: I'm not going to discuss it. I'm not going to discuss that.

INTERVIEWER: Can you help me understand why?

SCIENTIST: You will have to accept it. You have to accept it. Not me. (10L)

Yet, regardless of elites' post-career disappointment with achievements and professional standing, they are, as in earlier career phases, uniform in the way they assess their institutions: they remain havens, facilitators of their work, near-blessed places that permit them to continue, even in retirement, and, as always, free of "bureaucracy" and perceptions of associated entanglement. And, despite serious concerns about professional standing, all the elites in the subsample would seek an academic career again. Those who said they would do things differently referred to specifically technical decisions they had made in the course of scientific research that they now realize may have had consequences for subsequent achievement, a point that harkens back to associated patterns of career review and regret about how, and why, it has added up the way it has. The first of the passages below comes from the one woman scientist of the present cohort, and again highlights the correspondence between women scientists and elite career patterns.

[The place] is like drinking water out of a fire hose. That's the style. ... I am completely satisfied. I don't want to change a whole lot. If there's something I'm not happy with, I just say something about it, and it pretty much changes. I think that we have good rapport with our department heads. ... I don't have complaints. (14L)

INTERVIEWER: If you were starting all over again, knowing what you know now, would you go into an academic career?

SCIENTIST: Oh, yes. I was always able to decide what I wanted to do. [In doing things differently,] I'd take more mathematics, and I would try to have a bigger picture, to look around me, to see the opening where nobody's looking, and make the fantastic discovery. I could do it now. You have to work hard. You can't make those discoveries if you don't have the tools at your fingertips. (3L)

Oh, yes, [I would go into academia again] without any hesitation. I can't imagine a more satisfying career than the one I've had. I enjoyed teaching. I enjoyed research. I had good support. I chose the right university. I feel I was very lucky. I have no complaints at all. I'm very grateful for the generosity of [this institution] and the leadership to make it convenient for me to continue my work. (13L)

Two additional career characteristics fill in the picture of elites in latter- to post-career phases: estimations of their peak careers and their perception of reward systems. With regard to estimations of career peaks, elites in latter- to post-career phases, for the first time, identify not the present as the peak, but customarily turn to early-career phases to see themselves in their prime. Where once their prime was the ever-progressing present, up to the time when their careers reach their conclusions, now it is a specific point in the past, that is, associated with a scientist's sense of greatest promise. Moreover, at early career, scientists are situated in institutional conditions that would force their productivity to be comparatively high, and thus also, by sheer volume, more probable of receiving recognition. The change in identifying a career peak, from a long-standing present to a now-distant past, also indicates, among elites, a career about as fully formed as it will be. Soon it, and a life that constituted it, will end.

I was at a peak in 1962, maybe 1963, let's say '61 to '63. I published a series of papers on nuclear physics. They were recognized very quickly as being very good. I got lots of invitations to speak everywhere and to give courses in summer schools everywhere. I was in fashion. I was on the circuit. It continued for a long time. (20L)

[The peak] was when I was really deeply involved in the work which we were recognized for [in the form of a Nobel Prize], because that was so unbelievably exciting. It was from '67 to about '75 or so. That was probably the high point. We made discoveries later, but they weren't of the same magnitude

INTERVIEWER: Did you know at the time you were on to something?

SCIENTIST: Yes, yes. It became quite clear quite early. What it would turn out to be wasn't clear for quite a while, but we understood it was something big. There was an enormous amount of attention in those days to what we were doing. Everybody wanted to know what the latest measurements indicated. (16L)

Still, regarding the final of the career characteristics, elites — despite wide recognition — sense failure and lack of fairness in the operation of the reward system. This, too, marks a significant change. Once believers in the system, now they are more skeptical. Running short on time to continue exchange of scientific work with the system of reward, elites in latter-most career phases are now more apt to recognize, and be disappointed by, perceived shortchange.

[The reward system in science] *attempts* to be fair. There are many people who do wonderful things who don't get properly recognized. There's no question about that. I think I was lucky that the work we did was recognized. It could have been otherwise. I mean, after all, we got the Nobel Prize. We did the work probably between '67 and '74. We got the Nobel Prize [many years later]. It took close to twenty years to get that recognition. It could have been that we never would have gotten the recognition. We were very fortunate that it came our way because there is uncertainty in that kind of thing. There are many people who do wonderful things who don't get properly recognized. This last Nobel Prize — the three guys who got it for asymptotic freedom in the strong interactions — it took them thirty years, and there was a chance they wouldn't have gotten it. If it takes thirty years, it means that there was a chance that you won't get it, because something has prevented it [from occurring sooner]. (16L)

There are always some things that might be a little different and might be a little better. There are some areas where I thought I had some good ideas and good opportunities for research, from a scientific standpoint. I wasn't able to get funding, and this has happened over the years. I could complain about that, that they didn't fund me and they funded other people who I thought were doing a lot less. (14L)

I think I could get more credit for many good things I did. There are other people in the same condition.... It all has to do with recognition. There are certain things, things that I did years ago, that would be nice to get quoted. I could have gotten better recognition.... [I would be more fully satisfied] probably if I got... more recognition for my papers. (8L)

Were elites unconcerned about professional standing, one would not find them ruminating about papers written decades ago that they wish were cited or referring to grants that were submitted years earlier but went unfunded. Nor would they speak of other forms of recognition delayed, ungranted, or granted only in amounts that still leave them wanting. Professional standing rises as their prominent concern, and more so now than at any other time.

PLURALISTS

Whereas attenuation characterizes the career patterns of the most senior elites, *withdrawal* characterizes those of pluralists. The difference lies in the role of research in the career, again testifying to the prominence of research in defining much of the character that the career comes to possess. In attenuation, elites remain engaged in research, though less intensely. In withdrawal, research is left behind or substantially minimized. Of the most senior pluralists who have retired, few have remained active in research. While they keep offices in their department, they now function predominantly as hubs of leisure during periodic visitation. A minority of these retirees will carry on "light" research, as not to constitute a "program of" research, with multiyear goals and objectives that may lead to further lines of research, a qualitative distinction from their research in previous times. In keeping with prior patterns that underscored utilitarian outlooks, these pluralists are calculating their retirement: they know when they will retire and are waiting for the time to come. They envision a more leisurely engagement with scientific research, to the extent they envision this at all. They appear to enjoy more leisure.

In these respects, as in those presented in previous discussion, the pluralist world is accommodating. Pluralists view their department and institution no longer as a happy medium, even as there appears to be a comparatively high level of contentment, but more simply as a "place of work." It has become a place of habits in which a winding down of routines stands superior over actively seeking something new, including a career disposition toward major research or scientific breakthroughs. In all the cases, pluralists' work attitudes are positive and seem more positive than at prior points in time. There appear to be few, if any, major misgivings about retirement or plans to retire. For those who are not yet retired, the peak in the career is said to be now at its end.

> I am happier here at home, in my local institution, and I'm less involved at the national institutes [and meetings] outside of the university. Ten years ago, I was thinking that I would retire after thirty years, and that obviously didn't happen, because it's now been thirty-two years. That's partly because things are so much better here that, you know, it's fun. And there's a big financial incentive in our retirement system to spend thirty-five years. So I've readjusted my thinking because of that, to stay thirty-five years. . . . I will retire in three years. I've told all my colleagues that I will. We work on a three-year [grant] renewal. We just reviewed, and my reviews were very successful. I could probably go on. I told them I won't go in on the next grant, don't plan on me. . . . I will get the same salary when I retire as I will when I'm working. So I take no cut in salary, and I can continue to be as active as I like. Professionally, I'll still have my office. I'm very active in a [specialized retreat] for physics [in the West]. I will continue to be active there. I have a home [in the West]. I'll spend four or five months [in the West] in the winter and ski, as long as my knees hold out. I'll spend the fall here, being involved in research, helping other people on seminars and things like that, going to football games, playing golf. I will probably buy another home in some warm climate, maybe [in California], and I'll probably spend winter months [there], playing golf. Maybe I'll bug my friends at Caltech now and then, something like that. This winter I'm going to Florida for five weeks, just to try it out. I'm leaving on Saturday. I'll bug my friends in Gainesville, I'm close enough to drive in a day or two. I'll stay in touch with things here. . . . I'm very satisfied. I'm pretty happy here, things are going well. I worry more about whether my health will continue to be good and things like that. . . . Life is good.

INTERVIEWER: Has there been a particular period in your career or age where you think you have been most satisfied?

SCIENTIST: Yes, now. This is it. Yes, absolutely. There's no question about it.... The [departmental] leadership is better, the people I work with have been better. Also, I'm a little older, and I've had the opportunity to look back and see how great it has been over the years, to see the whole career collectively and appreciate how lucky I've been to do all the things I have done. That's a good feeling, and it's like, wow, this has been great. (50L)

Yet ten years previously, for many pluralists in middle to late phases, the career held less excitement; it is in anticipating its end where excitement and satisfaction appear to escalate. The scientist quoted above is quoted below at that prior point in time:

A number of years ago, I took a pretty realistic view of a career in theoretical physics. Theoretical physics is something that's done by young people, if you look at the major breakthroughs. They are done primarily by young people in the field. If you take a look around the community at what happens, you notice that careers, when people get into their sixties, tend to flatten out and usually turn down. I started planning in that direction a long time ago, with the point of view that I would probably be happier if I got out of this research in my early sixties and looked on to do something else.... I'm fifty-two, so I will probably retire in about nine years or so, maybe eight years.... I've gotten involved in other things. I have a condo [in the West] because I like the mountains. I like to ski, I like to hike and play tennis. (50F)

The withdrawal from work, and in particular scientific research, is captured further in the account of the following scientist, who, after a distinguished career, retired at the age of sixty-three. The withdrawal most characteristic of pluralists is set in tonal and substantive contrast to attenuation, discussed previously, as most characteristic of elites. Further associated with withdrawal are professional aspirations that continue to subside and the search for very little, if any, external recognition.

Around [the year] 2000 [five years after formally retiring], my career came to an end, I would say. I realized that I wasn't going to do anything terribly significant after that. Although I have published some papers, there are lots of things I could publish. Since about 2001, I would say I don't think about my career. Before then, one wanted to get a grant and wanted to continue to do really significant things. Now I don't think about it. (53L)

Or, as a colleague who retired in the same general period of time observed:

The amazing thing is, having retired in the year 2000, that I have had very little professional activity with physics. I'll go to the department to have lunch with people, to see people on personal terms, to keep up with things. But in terms of actually doing professional physics, even reading about my area, my life changed drastically when I retired. My attitude is kind of neutral or null. I don't do physics anymore. . . . I just don't think about it much. . . . I'm less hurried now and am happy to be away on the [hobby] subjects that I want to work on.
INTERVIEWER: . . . What worries or concerns do you have about your career?
SCIENTIST: I don't think about it. It never enters my head since working on [my hobbies].
INTERVIEWER: Are there things that get you down about your career?
SCIENTIST: As I say, I don't think about it. I don't go there [to that subject]. I have a new "career" [of other interests].
INTERVIEWER: Has the physics really gone away?
SCIENTIST: Like a bomb [laughter]. (54L)

Although retired pluralists appear to be at a kind of personal peak in retirement, in part, it would seem, because work is behind them, they identify their peak at work as at mid-career, constituting a disjuncture from their peers nearing retirement but who have yet to formally do so. Retirement appears to operate as an event that brings about a more complete review of a career, as to differentiate points at which scientists see themselves at their peaks.

I think the years when I was an associate professor were pretty productive. They were good years. That probably was when I was most satisfied. (49L)

When was I most creative? Probably in my forties. In my forties. I was terribly energetic and thought about physics all the time and worked very, very hard at it and had lots of ideas.

INTERVIEWER: Why, for example, your forties and not your thirties?

SCIENTIST: I did some very good things in my thirties, too. But by the age of forty, I had a fairly large group of students and postdocs and produced a lot of stuff. I was exploiting things I had gotten into in my thirties. (53L)

By late to post-career, the system of recognition, to pluralists, is now fair, where, to their younger counterparts, it was viewed as less than fair, at least by a major subset of those in earlier phases. Thus, as pluralists approach and enter retirement, their views of the ways by which scientific careers are rewarded and recognized become more forgiving. It is a striking association of patterns that their attitudes toward work, and life overall, grow more positive and also, as it turns out, that they now believe, on balance, that their careers progressed as expected. Measures of success become more internal and less external. A pluralist life in science, now gone or nearly over, grows more harmonious.

INTERVIEWER: Do you feel sufficiently recognized for your work?

SCIENTIST: Oh, yes. I would say yes.

INTERVIEWER: Do you think the system of recognition has been fair to you?

SCIENTIST: Yes.

INTERVIEWER: Fair and equitable?

SCIENTIST: Oh, yes. This system of recognition is not terribly important. There are these prizes and things like the Nobel Prize—they're not that meaningful. You probably know that Feynman refused to be in the National Academy because he said that they wasted most of their time considering who was good enough to join them in the Academy. (53L)

I feel the recognition is appropriate for the level I've achieved. I have a pretty good view of where I fit into the overall scheme of things, both with national research as well as within the university. I think it's close to a reasonable level of recognition. An academic career, you know my view of it, is whenever I allowed other people to set the measure of the reward, that could get me into a problem. I was always happier when

I set my own standards. . . . If I allowed myself, and it's easy to say, to think, "If I do a really good job and I get nominated for dah, dah, dah, I'll get a big raise," I was setting myself up for a bad situation. It's easy to allow yourself to get into that pattern. The system is constantly oriented that way. If you do that, you're setting yourself up for disappointment. (50L)

In general, pluralists in late to post-career would entertain the possibility of an academic career again; a minority would do so without hesitation. While their attitudes toward the system of reward have grown more positive, they still recognize difficulties with academic careers, such as the availability of academic employment, the challenge of ascending academic ranks, and the remote chances of major scientific discovery. Some pluralists indicated that, while they would probably seek an academic career again, they might select a field other than physics.

INTERVIEWER: If you were starting all over again, knowing what you know now, would you go into an academic career?
SCIENTIST: I don't know. I really don't know. Academic careers are certainly not as attractive as they were when I was starting out [in the mid-1960s]. I don't know. It's not a definite "yes," but it certainly isn't a "no." I can well imagine that there would be other careers out there, about which I don't know much, that I probably would prefer if I knew more about them.
INTERVIEWER: If you were to go into an academic career again, what do you think you would do differently?
SCIENTIST: I might not go into physics. I might have gone into computer science—it didn't exist as a major when I was an undergraduate. (49L)

INTERVIEWER: If you were starting all over again, knowing what you know now about academic careers, would you go into one again?
SCIENTIST: Oh, yes. I think there's no question that it's the best thing there is. It's so varied. It's as though you're working for yourself, most of the time. You decide what you're going to do, and you look forward to doing it, and you enjoy doing it. It's just great. It's pure fun, most of it.
INTERVIEWER: If you were starting all over again, knowing what you know now, what would you do differently?

SCIENTIST: I probably wouldn't go into the particular field of physics [low-temperature physics] that I did go into, because, well, it's a hell of a long time since then, and physics has changed and made a tremendous amount of progress. I might even go into some other field, not physics. [Possibly] genetic engineering, or something like that. Something involving DNA and genes and things like that. That's terribly exciting now.
INTERVIEWER: And the rationale for shifting the specialty is?
SCIENTIST: The opportunity to find out wonderful and fascinating things.
INTERVIEWER: These other fields you would characterize as less mature?
SCIENTIST: Yes. Absolutely.
INTERVIEWER: And therefore more open to discovery?
SCIENTIST: Yes. And also there are all these new techniques. They're doing experiments in such a wide range of opportunities. (53L)

INTERVIEWER: Would you recommend an academic career to someone else?
SCIENTIST: I would warn them about the difficulty. I have one star daughter, as an undergraduate, and a strong student with a PhD in English from [the University of] Michigan, and she's struggling getting an academic job. I saw that [in physics] along the way. You don't tell children what to do, but as you know, it's like all those sopranos who want to sing at the Met, or even sing at some regional company. Many of them don't sing at all. I would warn any person going into academia that that's the case. Most of your students going into sociology . . . probably think that they will get research jobs at major institutions, and you know that most of them are wrong. It's not a definite ["yes" that I would choose an academic career again]. I might choose another career. (54L)

The accounts of pluralists at these points in their careers are distinct from those of elites. Elites, overall, maintain their love affair with science. They become less satisfied because expectations have exceeded rewards. They question, for the first time in latter career phases, the fairness of how the scientific reward system operates. Pluralists, if they do not break their ties to science, substantially weaken them, marking their withdrawal. They become more satisfied. For the first time with any generality, they concede that the reward system of science is fair. Less tied to it, or out of it altogether, they have accepted their lot. Free from science, or on the cusp of retirement, they might even seek an academic career again, where earlier they met with greater frustration. The hapless bureaucracy that, once in their eyes, was their institution, lives on, but not in them. They have a place to work and a place from which to retire.

COMMUNITARIANS

Careers in the communitarian world of science undergo the greatest modifications. It is the world where scientists most greatly reduce, redirect, and redesign their professional aspirations. As earlier chapters have revealed, communitarians rework accounts of their careers over a significant fraction of them; much of the career account consists of explaining and justifying career outcomes. One might therefore expect communitarians in the latter-most phases of their careers to look back with considerable regret, to continue to grapple with the ways in which they see their careers not having worked out, to possibly be bitter and disappointed. The data reveal, however, that these expectations are dashed. In yet another illustration of reversal, communitarians in late- to post-career phases find a satisfaction that had alluded them at earlier points in their careers. The career is now seen to have gone "as expected"—it all comes to work out in the end. How is this accomplished? What are the associated patterns that characterize communitarians at the end of their careers?

In the present phases of their careers, communitarians have no apparent prominent concerns. The object of their greatest satisfaction is retirement. They display detached attitudes about work. They say they work less hard or are not working at all. Possessing exclusively internal definitions of success, whenever necessary to utilize them, they no longer seek professional recognition and have no professional aspirations. They have essentially removed themselves from work, an orientation that is as characteristic for those who have formally retired as for those soon expecting to retire. The overall career pattern consists in separation, whereby individual and institution, once intricately bound, undergo a peaceable and permanent divorce. No longer viewed so much as "stymieing," their departments and institutions are now "a place departed." Free of place, in mind if not yet in action, communitarians achieve relatively high levels of satisfaction.

INTERVIEWER: Do you miss teaching?

SCIENTIST: No, no. Not now. I decided to retire, and once you make that decision — I had plenty of time to think about it. I had plenty of time to just wind down and to develop a different perspective and different attitude toward things, especially realizing that people don't really want to hear what I've got to say anyway, so why bother? . . . [My outlook] is tempered by the realization that I haven't been all that successful in science. Most people aren't Einsteins by any stretch. One goes in expecting to make great discoveries and to participate in a great understanding of things. The reality is, with most people, they end up doing less than they would have liked about being able to solve big problems. The realization that the important work that one would have liked to have done is not going to happen — it dampens one's enthusiasm. . . . I might have contributed more to society by going in a different field. I might have been able to do more. I might have been better working on technical problems, perhaps as an engineer or something.

INTERVIEWER: How often do you find yourself thinking about your career?

SCIENTIST: I don't anymore. I don't have to worry about that. . . . [The best part of retirement is] being free. Not having constraints of work. Not having to show up at a given time and do certain things. Just being able to do whatever you want, whenever you want to do it. Just not being tied down anywhere. . . . There haven't been any bad parts.

INTERVIEWER: Have there been any adjustments you have had to make?

SCIENTIST: Surprisingly, no.

INTERVIEWER: Has it been any problem in organizing your days?

SCIENTIST: No. There are so many things to do. (31L)

I was tired of teaching. The quality of our . . . students was declining. I just had less patience for dealing with students. . . . So I decided it wasn't worth the aggravation. So I did it. I travel a lot.

INTERVIEWER: How would you complete the sentence, "I am more X and less Y compared to ten years ago"?

SCIENTIST: I'm more calm and less frustrated. I think I had been frustrated for quite a while and just sublimated it. And I was disappointed, because I thought we had done some good work. I thought we turned out some pretty good papers and some interesting ideas. We couldn't get [grant] support to go on. The interest in what we were doing just didn't stay.

INTERVIEWER: . . . Do you think your career progressed, up until the time you retired, as you had wanted it to?

SCIENTIST: Pretty much so, yes. . . . There really wasn't much else to look forward to. [Right now, I'm] not working as hard. I'm not doing research anymore. I had two or three pretty good ideas during the course of my career, and I haven't had any since. I really don't keep up with the literature. . . . I think early on, even though I did some fairly decent work, both as a graduate student and in the beginning of my career, I never was satisfied. I always thought that I could have done better or sooner or more. In more recent years, I have become content, not only with what I was doing, but also how much. I think this is a reflection of my coming to like myself more.

INTERVIEWER: What worries or concerns would you say you have about your career?

SCIENTIST: None, now. My career as a physicist is over. I talk to colleagues occasionally. But they've gone on to do other things. The faculty has either retired or has gone on to try to do something else.

INTERVIEWER: . . . Since retiring, what do you miss most about your job?

SCIENTIST: Not a hell of a lot.

INTERVIEWER: Is there anything that you miss?

SCIENTIST: No. Not at all. There are very few people that I really enjoy being around, and none of them are my former colleagues. I find them boring. This one guy was a very, is still a good friend. But, you know, I'm around him for fifteen, twenty minutes, and I'm thinking, I've got to get away. He rattles on and on about the same old things.

INTERVIEWER: . . . What has been the best part of retirement?

SCIENTIST: Doing whatever the hell I want. I can get up and go to the [gym] and work out, or ride my bike [downtown] and have coffee, or even go over to the department—I don't do that very much anymore. (35L)

More so than at any other time, success is defined internally; external measures and arbiters of success are rendered irrelevant, elevating a detached attitude toward work.

INTERVIEWER: Do you think you define success differently now than you did in the years prior to retirement?

SCIENTIST: I don't really care about that stuff anymore. I'm successful everyday if I do what I was hoping to do that day. My success is a little bit easier for me to control and define. I'm pretty happy. My happiness is defined totally by me, not by anybody else.

INTERVIEWER: What would you say are your current aspirations?

SCIENTIST: My biggest aspiration is to stay healthy and happy. And enjoy my life.

INTERVIEWER: What do you miss the most about your job?

SCIENTIST: I don't miss anything about my job. That's one of the biggest surprises to me.... I kept working my butt off. As soon as I made the decision [to retire], I started to wonder about how I was going to feel. I can still remember the day I was sitting here, during the Christmas holiday, where I live now, filling out the papers. I was asking myself questions like, "How are you going to feel if you do that?" And I did it. And you know something? I have never thought about it again. I am feeling great. It felt really good to think about something else. There are other things in life to think about than the things I had been thinking about. (28L)

There is irony in the separation communitarians make from their world. The communitarian world is built and maintained by people who, in their careers, must turn to one another locally in order to derive meaning about their work and performance in it, in part because the more cosmopolitan arenas of research are unavailable due to shortage of opportunity and in part because of the real occupational demands in place at these institutions. It is a world in which immediate activities, teaching and service, are assigned great value, because they serve substantial functions, further localizing a system of status and the meanings that people are able to derive from their careers. At the end of the career, this localized world is rejected. The localized activities, teaching and service, are surrendered with relief. The colleagues one has known for decades are now cast away, seen rarely, and often avoided. It provides another illustration of the instability of careers, of academic environments, and of the propensity for individual adaptation, when research fails to be a strong institutional component of academic organization.

The passages above, as with the passages of other communitarians at the end of their careers, call attention to an additional prominent characteristic that distinguishes these individuals from others, especially scientists in the elite world. Notice the ease not only with which communitarian scientists are able to discuss their perceived failures, but also how readily they dispense with them. Perceived shortcomings in research may be acknowledged by the most senior communitarian, but easily put aside and forgotten. It is not viewed as a threatening subject for discussion. Whereas for the most senior elites, quite the contrary was found. Not only were they uneasy talking about perceived failures in research, they could grow angry about it. Professional standing was the prominent concern by the end of the career. In reducing expectations for superior performance earlier in their careers, communitarians attain a more forgiving sense of self in the end. In maintaining heightened career expectations, elites, paradoxically in the end, confront a greater sense of disappointment.

Communitarians are apt to identify their peak satisfaction at two locations. One of them is the present, in or near retirement.

> INTERVIEWER: At what point in your career do you think you were most satisfied?
> SCIENTIST: Maybe now, because my career is coming to an end. (26L)

Another scientist, recently retired, put the matter into further perspective:

> I think the worst thing that one could possibly do is live on expectations.... Your disappointments that you have in life are directly proportional to your expectations. Having expectations is what leads to disappointment.... I feel okay about what I did [in my career].... [I] changed my outlook from worrying about what was going to happen, and I became happy with what I had, rather than worrying about what I didn't have.... I'm totally unconcerned. There's a certain kind of ease that I didn't always have, that I have now. I feel like everything is going the way it should. I'm not worrying about what I need to do next or what I need to do to succeed. (28L)

The second location is the early career. Early career marks a time of substantial optimism and promise about future achievements. Career expectations have yet to be reset. In early career, communitarians were at their most productive, and thus likely the most recognized in the span of their careers, at least for those contributions that were received relatively quickly.

[I was most satisfied] probably when I was younger and running from place to place, always at high speed. [It was when] I started out, when I was doing my research, and I was pretty enthusiastic about all that stuff. (31L)

While, by the end of their careers, communitarians have found new satisfaction, and while they typically reconstruct their careers as having gone as expected, they view the reward system of science as unfair overall. But, in keeping with the new life they have found, they tend to emphasize the rewards of retirement from their profession, actual or anticipated, rather than perceived failings of their professional reward system.

> INTERVIEWER: Do you feel sufficiently recognized for your work?
> SCIENTIST: Probably not. . . . I thought I should have been elected a fellow, not of physicists, but of geophysicists, because that's where I've published some really good stuff. (35L)

I'd like to feel more recognized. . . . There was a period of time, from '75 to '85, where I felt like I was involved in some professional controversies about some ideas that a collaborator and I had developed and published, and there was some opposition to those ideas—professional controversy. But I feel we were vindicated by future events—these ideas became acceptable. . . . I'd like to be more recognized. (26L)

Despite the satisfaction they have been able to find at the close of their careers, they understand it as hard-won satisfaction. Most communitarians in these career phases are doubtful that they would seek an academic career again. If they were, many would seek a different type of university or a different field, in each case pointing in a direction where they perceive greater professional opportunity.

> INTERVIEWER: Would you go into an academic career if you were starting all over again?
> SCIENTIST: Probably not. It's very, very difficult today. There are a lot of demands. You are not left alone to do your work. There's a great deal of pressure on young people to get large grants. And they're also expected to do a superb job in the classroom. It's a difficult life for the younger people. I'm not sure I would choose to go into physics. I'm

looking at it from my perspective today. I think it's an extremely difficult field and demands a lot of time. I'm not sure I'd be willing to do that. It has too much of a personal commitment, and the rewards just aren't there. (40L)

I would tell any young [person] getting ready to start a career in the university to think very carefully about it. Don't have terribly high expectations as far as rewards. It's a different climate. . . . [I might] go into a stronger department . . . or university . . . it was hard working for everything [all the resources necessary for work]. It was a distraction. We had to fight to get [resources] . . . it was a tremendous battle. There were a lot of casualties along the way. I thought about leaving, myself, at one time, but I didn't. (35L)

I'm not sure that I really would [seek an academic career again]. In some ways, I don't think it is really worth it. The rewards that you get and the pressures that would be on the person to generate all these funds, and if you don't, you're not going to be around for very long. . . . I think it's harder now. There's so much more competition. More is expected of the younger people now. . . . I think I would try to do something else, to contribute something useful to society, and I might be able to do something better in another area. [If it were an academic career,] I would start off in a different field. (31L)

SUMMARY

By the end of their careers, scientists across the academic worlds have come to view themselves in relation to their work in often radically distinct ways. Contrasts between elites and communitarians, at either end of the continuum of academic worlds, are especially pronounced. As observed, several of these differences reflect a deepening of patterns that originated in previous decades. But it is also apparent how other of these differences express themselves for the first time at the conclusion of scientists' careers, in many instances constituting a reversal of outlook and perception evinced at earlier points in time. Table 28 provides a summary of the patterns that characterize scientists in their passages to later/post-career phases. The table is parallel in structure to tables 22 and 25, which provided a summary of career patterns for the younger two cohorts of scientists.

While elites continue to focus their careers on research, they do so with

TABLE 28. Late- to later/post-career patterns of scientists

Career Dimensions	Elites	Pluralists	Communitarians
CAREER FOCUS			
In Late Career	Research	Research & Teaching/Teaching	Teaching
At Later/Post-Career	Research	Light Research Not Professionally Active	Not Professionally Active
PROFESSIONAL ASPIRATIONS			
In Late Career	Remain Steady	Subside	Not Present
At Later/Post-Career	Tempered	Subside	Not Present
RECOGNITION SOUGHT			
In Late Career	Great	Average	None
At Later/Post-Career	Happenstance	Low; None	None
ORIENTATION TO WORK			
In Late Career	Moral	Moral or Utilitarian	Utilitarian
At Later/Post-Career	Intellectual	Utilitarian	Removed
WORK/FAMILY FOCUS			
In Late Career	Work	Family & Work	Family & Leisure
At Later/Post-Career	Work & Leisure	Leisure & Work; Leisure	Leisure
ATTRIBUTION OF PLACE			
In Late Career	"Haven"	"Happy Medium"	"Stymieing"
At Later/Post-Career	"Haven"	"Place of Work"	"Place Departed"
OVERALL SATISFACTION			
In Late Career	High	Medium	Low
At Later/Post-Career	Medium-Low	High	Medium-High

(continues)

TABLE 28. (continued)

	Additional Later/Post-Career Patterns		
Career Dimensions	Elites	Pluralists	Communitarians
CAREER PROGRESS	Not as Expected	As Expected	As Expected
WORK INTENSITY	As Hard; Less Hard	Less Hard	Less Hard
OBJECT OF SATISFACTION	Research	Light Research; Retirement	Retirement
PEAK SATISFACTION	Early Career	Present; Mid-Career	Present; Early Career
REWARD SYSTEM	Faulty	Fair	Unfair
DEFINITION OF SUCCESS	External	Primarily Internal	Internal
WORK ATTITUDE	Ambivalent	Positive	Detached
PROMINENT CONCERNS	Professional Standing	Retirement Timing	None
ACADEMIC CAREER AGAIN	Yes	Maybe; Yes	No
DO DIFFERENTLY	Technical Decisions	Different Field	Different Field/ Different Institution
LEAVE UNIVERSITY	No	No	No
RETIRE NOW	Retired; Planning	Retired; Planning	Retired; Planning
OVERALL MODAL PATTERN	Attenuation	Withdrawal	Separation

more tempered professional aspirations, in which they adopt a more freely intellectual orientation to work. Though they would look kindly upon additional recognition, this is not what primarily motivates their continued research involvement in these phases. What recognition they would receive is viewed as happenstance. Along these lines, one finds them professing that they work either as hard or less hard than previously, and somewhat more ready to discuss leisure in conjunction with their professional roles. They continue to regard their departments and institutions as havens for allowing them, in turn, to continue in their most valued activity. Overall, their career patterns display *attenuation*—a less intense engagement with science, academia, and their careers.

Among the most striking developments among elites at the end of their careers is their change in work attitude, which, while moderately positive, becomes ambivalent. At the end, elites develop a heightened concern about their professional standing as they realize the improbability of change in the standing they have achieved. In a reversal, their overall satisfaction drops, best described as "medium-low," that reflects a disappointment in not having achieved more. In face-to-face interviews, elites solemnly communicate this void as a kind of dull, prolonged ache—as if grieving a loss—and at other times in an angry bitterness. Among all the scientists at the end of their careers, it was most difficult to talk with elites, arguably the highest achieving, about achievement. In another twist, elites—the most rewarded scientists—perceive their careers as not having gone as expected.

A major cause of these sentiments lies in how elites define success, even at the latter-most career phases: external definitions, tied to the reward system of science, are invoked to establish standing and status. Because many elites see themselves falling short by the end of their careers, they are most apt in this period to identify faults with the scientific reward system, constituting yet another reversal in attitude compared with previous points in time. They invoke early career as their peak, a time that corresponds to promise and productivity as well as recognition, received and anticipated. Despite serious misgivings, the eldest elites would seek an academic career again but would look more carefully at technical decisions that are made in the course of scientific research, which they now, more than ever, understand to have fateful personal outcomes.

Pluralists *withdraw* from work, including, for most, a stoppage of research. Those who continue in research do so lightly. On the whole, they have calculated when to retire, contributing to a utilitarian outlook on work. They seek minimal or no recognition; their professional aspirations subside, and one hears of the more prominent place that leisure holds in their accounting of the routines characterizing them at this point in time. Where once a "happy medium," in which teaching and research existed side by side, perhaps one of these roles stressed over the other, perhaps only in one phase and then reconfigured in another, the department and university are now simply "places of work." They house accustomed routines, and the individuals daily performing them, until retirement comes.

Previously fitful about ways in which their institutions constrained their careers, pluralists now see their careers as having gone as expected. Their overall work attitudes are positive; they are upbeat; they have few concerns, the timing of retirement most prominent among them. In another example of reversal, their peak satisfaction is the present, at least for those not yet retired, underscoring further the beneficence of time. Those who have ritually crossed into retirement engage in a more extended review and identify prior points, especially mid-career, as their peak, times associated with heightened productivity and promise of attainment in a career that has withstood early adjustments and achieved a thriving maturity.

These positive outlooks are associated with a perception of the scientific reward system as fair in general, though this is coupled with the finding that pluralists in these phases now adopt primarily internal definitions of success. They have left the reward system of science for others, have reconciled themselves with it, and are settled comfortably with their own standards of success. In general, they would consider an academic career again, though not necessarily in physics, further partial testimony to the greater positive light in which the career is viewed when it is over or nearly over.

Whereas pluralists withdraw, communitarians *separate* from work, science, and their institutions altogether. Work and institution are viewed as "a place departed": individuals cease to identify with their departments or universities or, for the most part, with science. Individual and institution become separate entities. Those who have retired, and even those who have not, are no longer professionally active. They do not seek professional recognition, nor do they discuss or convey professional aspirations. Communitarians have removed themselves from work, even if they are still there, waiting to retire. Above all else, they stress the significance of leisure activities in the constellation of their activities.

Retirement is the primary object of their satisfaction, and for the first time in their careers, they have no prominent concerns about work, toward which they now exhibit an altogether detached attitude. In yet another instance of reversal in the end, communitarians, almost miraculously, state that they see their careers as having gone as expected. No longer, or not much longer, a part of science or academe, it all comes to work out in the end, highlighting—as with pluralists—how fine work and career seem once gone. It is as if retirement works a kind of magic on perception.

Communitarians are apt to identify either their early careers or the present as their peak. As with others, early careers recall a time of productivity and perceived promise. The present is a time of newfound liberation. They regard the reward system of science as unfair overall but have come to accept it, in part by having adopted internal definitions of success that are more forgiving. In contrast to elites and pluralists, communitarians by the end, in wide agreement, would not seek an academic career again and, if compelled, would pursue a different field and a different type of university where they would hope professional opportunity would be more abundant. These sentiments notwithstanding, communitarians are at their highest in overall satisfaction, which may best be described as "medium-high," in light of all considered. Now that their careers have come to a conclusion, communitarians find a kind of contentment that previously eluded them.

CHAPTER FIVE
Lives of Learning

What does the academic profession mean to the people who comprise it? To be certain, it involves lengthy education and technical training, which have grown lengthier and more technical as knowledge has evolved and grown more specialized. It includes theoretic knowledge and jurisdictional control, even as the content and scope have changed over time. Access is restricted, although higher learning, including graduate and professional education, is now more accessible than before. Its members enjoy an autonomy, although arguably not as much as in earlier eras (for a discussion of these and related changes, see Rhoades 1998). Academia is made up, in varying degrees, of these and other classic traits known to constitute professions. But to the *people* of it, academia, and specifically science, consists of subjective careers that unfold differentially across settings in which they work. The evidence also indicates that the meaning of science varies at different times in scientists' careers. This work has viewed academia through a different lens: I have approached the study of a profession, not by virtue of its various structural traits per se, but in terms of unfolding careers. In allegedly one profession, one sees many highly differentiated careers.

In the preceding chapters, I have examined careers in science by way of three points of intellectual departure: a perspective on occupations that has stressed the reciprocity between individuals and institutions in each other's creation and identity substance; a perspective on the life course that has stressed aging within cohorts in order to reveal differences in career outlook within and across phases of time; and a perspective on the sociology of science that has stressed strata of scientific practice and experience of the scientific life within the strata. I have brought these perspectives together to form the basis of this inquiry into how scientists, in varieties of academic contexts, perceive their work over time.

In this chapter, the discussion turns to accounting for the career patterns observed across the contexts of science and cohorts of scientists. I consider the following questions: What happens to careers in science? How do outlooks and orientations to work change? Why do these changes occur? And how can these changes be explained? Answers to these questions invariably give rise to others, which I shall also address, including: Are academic careers in science worthwhile? And how have the rewards of academic careers changed? Finally, I will consider the generalizability of the present findings to members of other scientific and academic fields as well as what the future may hold for academic careers in science.

In sections of the discussion that follows, I draw suggestive conclusions based upon the study's main findings and patterns in the data. In general, these conclusions are framed as propositions. Accordingly, they may be used as bases for further inquiry into the study of academic careers, the academic profession, academic fields, departments, and universities. While several findings are covered, including those represented in tables that appear both in this and in preceding chapters, the discussion highlights forty propositions organized into thirty sets. The propositions are notated in *italics*; they have also been consolidated for summary and reference in appendix G.

EXPECTATIONS AND THE RHYTHM OF CAREERS

As scientists age, they evince numerous changes in their identification with work. In the three previous chapters, I have examined the ways in which scientists identify with their careers vis-à-vis the career dimensions (displayed in tables 22, 25, and 28) that situate their experience and understanding of work. At the outset, one might have expected that elites would find their work the most rewarding, that their satisfaction would also intensify over time such that, given the most accomplishment, senior elites would be easiest with whom to discuss professional achievement. Conversely, one might have expected that communitarians would find their work the least satisfying, that this lack of satisfaction, too, would intensify over time such that, given the comparatively meager opportunity to accomplish, senior communitarians would be the most difficult with whom to discuss professional achievement. One might have expected pluralists to exemplify a mix of these patterns, given that they are structurally and culturally situated in the middle of the institutional continuum.

Moreover, one might have expected that elites would be the most satisfied in their work since they are the most recognized and publicly validated. Correspondingly, eldest elites would be found to exhibit especially high levels of satisfaction in their careers. By contrast, one might have expected communitarians to be least satisfied in their work since, comparatively, they are the least recognized and publicly validated. Correspondingly, eldest communitarians would be found to exhibit especially low levels of satisfaction in their careers because they have endured the longest period of fewer rewards and recognition for their efforts. One might have expected pluralists to exemplify a mix of these patterns in light of their intermediate structural and cultural location.

Further still, one might have expected that elites would be more inclined to believe that the reward system of science is fair, and have confidence that they have been fairly recognized, because, comparatively, they are the most invested in and successful at scientific research. Conversely, one might have expected that communitarians would be least inclined to believe that the reward system of science is fair and have little confidence that they have been fairly recognized, in light of their comparatively modest engagement in scientific research. Of pluralists, one might have expected a combination of these patterns.

The cross-sectional work reported in the *foundational study* of scientists' careers could have led easily to these claims. To what stage had scientists proceeded by that point in time? Ten years earlier in their careers, before the respective cohorts had progressed into mid-, late-, and post-career phases, the groups of scientists were confronting distinctive sets of career constraints and contingencies on their academic worlds. Scientists in early career alternately viewed their world as a "burden" (elites), a "happy medium" (pluralists), and as "stymieing" (communitarians). A generation ahead of them were scientists in late career, but well short of the ends of their careers, who alternately found high (elites), medium (pluralists), and low (communitarians) satisfaction in their work. Extrapolating forward in time, one would miss altogether the transformative phases in scientists' mid- and late careers and the empirical realities in which they would find themselves in retirement.

Across the three cohorts of scientists, however, the longitudinal data lead to strikingly different conclusions. The overall modal patterns of scientists' careers across the three settings of academic science, moving from early to mid-, mid- to late, and finally late to post-career, are presented in the first panel of table 29. Overall satisfaction by career phases is presented in the second panel, and work attitudes of the scientists are presented in the third panel of the table. *One observes notable reversals in outlook and identification with the career. In broad terms, elites enter mid-career highly satisfied, only to end them with ambivalence. Communitarians enter mid-career highly dissatisfied and end them*

TABLE 29. Modal career patterns, overall satisfaction, and work attitudes of scientists, by career phases

OVERALL MODAL CAREER PATTERNS OF SCIENTISTS

Phases	Elites	Pluralists	Communitarians
Early to Mid	Stabilization & Rededication	Reversal	Stasis
Mid to Late	Continuation	Regeneration or Continuation	Demise
Late to Post	Attenuation	Withdrawal	Separation

OVERALL SATISFACTION OF SCIENTISTS

Early	Medium	High	Low
Mid	High	Low	Low
Mid	High	Medium	Low
Late	High	Medium	Low
Late	High	Medium	Low
Post	Medium-Low	High	Medium-High

WORK ATTITUDES OF SCIENTISTS

Early to Mid	Positive	Preponderantly Negative	Preponderantly Negative
Mid to Late	Positive Positive	Ambivalent;	Neutralized
Late to Post	Ambivalent	Positive	Detached

with serenity. In the middle, pluralists start on a "high," proceed to either a low or moderate level of satisfaction, and conclude on another "high" (proposition 1). In their examination of professors at liberal arts colleges, where teaching predominates among professional roles, Roger Baldwin and Robert Blackburn observed similar patterns. Professors in late career, but still outside of five years of retirement, experienced reduced enthusiasm for research as well as teaching, questioned the value of their career activities, and sometimes the value of an academic career altogether. But within five years of retirement, professors were, by turn, content with their career achievements. They also had withdrawn from many of their professional responsibilities (Baldwin and Blackburn 1981; for related work, Bentley and Blackburn 1990; Blackburn and Lawrence 1986; Lawrence and Blackburn 1985).

By correspondence, attenuation in the latter-most phases of elites' careers is associated with lower levels of professional satisfaction and the development of an ambivalent work attitude. Withdrawal and separation in the latter-most phases of pluralists' and communitarians' respective careers is associated with higher levels of satisfaction (though not necessarily professional satisfaction), a positive work attitude among pluralists, and a detached attitude about work among communitarians (proposition 2). One is thus drawn to ask: What happens to scientists' professional careers over time? Why and how are these observed patterns produced?

To answer these questions, I turn to scientists' *expectations* for their careers. Sociological theory, most centrally a line of thought that extends from Emile Durkheim to Robert Merton, informs the dynamics of scientists' career expectations and how they are associated with the other career patterns observed in scientists' perspectives and attitudes over time.

ANOMIE AND ADAPTATION

In his classic work on suicide, Durkheim uncovered four principal forms of suicide, anomic suicide one of them, in which the form expresses a gap between individuals' expectations for the future and the realities of their present situations. Anomie (also *anomy*), strictly translated as "without law," is construed sociologically as a state of normlessness in which individuals suffer a breakdown of order, a collapse of meaning about themselves, the world, and their perceived place in it because of a sharp divide between the realities of their lives and the needs and wants for their future. Marshalling empirical evidence on economic conditions, Durkheim observed:

> No living being can be happy or even exist unless his needs are sufficiently proportioned to his means. In other words, if his needs require more than can be granted, or even merely something of a different sort, they will be under continual friction and can only function painfully. . . . To pursue a goal which is by definition unattainable is to condemn oneself to a state of perpetual unhappiness. . . . Thus, the more one has, the more one wants, since satisfactions received only stimulate instead of filling needs, . . . Our thread of life on these conditions is pretty thin, breakable at any instant. . . . Overweening ambition always exceeds the results obtained, great as they may be, since there is no

warning to pause here. Nothing gives satisfaction and all this agitation is uninterruptedly maintained without appeasement. Above all, since this race for an unattainable goal can give no other pleasure but that of the race itself, if it is one, once it is interrupted the participants are left empty-handed. (Durkheim [1897] 1951, 246–248, 253)

Durkheim observed changes in economic conditions of societies that brought both prosperity in some instances and peril in others. He used such alternate cases to demonstrate that it was not financial despair that caused anomie, since the same outcome could be observed arising from financial windfall. Rather, it was crisis in the collective order, brought about by downward or upward change, that established conditions for anomic suicide. For the modern-day Durkheim, the devastating loss suffered by the gambling hand establishes conditions like that of the hand holding the prized lottery ticket—both constitute a marked change in order and foster individual crisis because of the discrepancy between present reality and an anticipated future life.

For these reasons, Durkheim remarked on the seemingly paradoxical power of poverty. The "remarkable power of poor countries," according to Durkheim's theoretical stand, lies in the prevention of individual expectations that well exceed a capacity to realize them.

Poverty protects against suicide because it is a restraint in itself. No matter how one acts, desires have to depend upon resources to some extent; actual possessions are partly the criterion of those aspired to. So the less one has the less he is tempted to extend the range of his needs indefinitely. Lack of power, compelling moderation, accustoms men to it, while nothing excites envy if no one has superfluity. Wealth, on the other hand, by the power it bestows, deceives us into believing that we depend on ourselves only. Reducing the resistance we encounter from objects, it suggests the possibility of unlimited success against them. The less limited one feels, the more intolerable all limitation appears. Not without reason, therefore, have so many religions dwelt on the advantages and moral value of poverty. It is actually the best school for teaching self-restraint. (Durkheim [1897] 1951, 253–254)

One is able to see parallel conditions in modern science. Building upon Warren Hagstrom's early theorizing, "Anomy in science can be specified as the general absence of opportunities to achieve recognition" (Hagstrom

1965, 228). In the present instance, finding wealth of professional opportunity and experiencing cumulative advantages that further fuels performance, elites successively heighten expectations for themselves and their careers. Achievement only brings about desire for more achievement. At the end of the career, when greater achievement proves elusive, if only because of lack of time, but often also because of failed abilities and capacities, elites experience a reversal, develop an ambivalence about work, deem their careers not to have progressed as expected, and find fault with the system of scientific reward that so vigorously directed their efforts over the preceding decades. They have, in Durkheim's words, found themselves "empty-handed," even as their hands are full indeed. Expectations for their careers exceed reality. The comparison is, of course, analytic. Thus this is not to say that elites are prone to suicide but that, under the conditions wherein expectations for the future exceed opportunities to satisfy them, elites suffer from self-altering feelings of fragmentation.

At the end of their careers, elites customarily experience the phenomenon known as anomie. Communitarians and pluralists experience anomie also, but typically in much earlier phases of their careers, when it is possible for scientists in these worlds of science to realize that their career expectations cannot be realized (proposition 3a). Over many years, beginning in early- and extending into mid-career phases, such expectations are abandoned or significantly modified. This process normally occurs with substantial personal agony and feelings of professional loss, especially among communitarians. But by the end of their careers, communitarians are "serene." They are at the greatest peace with themselves and their careers than at any previous time, having achieved a detached attitude from work. While readily able to see faults and failings of the scientific reward system, they no longer care about it. By the end, they even see their careers as having progressed "as expected." In Durkheim's terms, "poverty" of their academic world, in the end, protects them. Working in a "fixed society," with limited opportunity for recognition and advancement beyond those of the ordinary academic ranks, communitarians develop limited aspirations because they know how much they can achieve. Lack of reward and recognition in this academic world thus functions as an effective self-restraint on "limitless aspiration" and vaunted expectations for self and career. Following Durkheim, one may conclude that the incidence and longevity of anomie among elites is greatest because elites are exposed to the greatest potential for rewards (proposition 3b). In this sense, substantial recognition is never enough for elites, even at the completion and in years following a high-achieving career. For communitarians, abundant recognition, and even modest levels of recognition, become irrelevant well before the conclusion of their careers. In principle, some pluralists exemplify the elite-tending pattern, others the communitarian-tending pattern. In practice, most pluralists by the end of their careers substantially withdraw career commitment and release their career expectations much like communitarians, in large measure because a career in a pluralist institution can never be as affirmed as those in elite institutions. Hence one observed the overall satisfaction and retrospective work attitudes of pluralists rising to a high in post-career.

Hagstrom construed anomie as a condition of the marginal scientist, occurring especially in highly arcane areas in which researchers are comparatively independent of others and thus at greatest risk of having their work go unnoticed. Hagstrom as well surmised that anomie may have been more prevalent in early rather than in modern science because norms governing recognition were relatively weak and loosely formulated (Hagstrom 1965). The present work derives contrasting results. One observes anomie throughout academic science, in both marginal and in integrated researchers. Important to note, anomie not only has professional, but also organizational bases. It is manifest differentially across organizational environments of academic work, arising at relatively early career points in some academic worlds (especially communitarian and pluralist) and relatively later career points in other academic worlds (especially the elite). This evidence also suggests that *anomie is more likely to arise when norms governing recognition are well developed, as in modern science, wherein the expectations to achieve recognition—transmitted organizationally through heightened university expectations in individual role performance—become pronounced* (proposition 3c).

These findings are partly consistent with Arne Kalleberg's general theory of job satisfaction. Kalleberg suggested that "the extent to which workers are able to obtain perceived job rewards is conceptualized to be a function of their degree of control over their employment situations" (Kalleberg 1977, 124). He further hypothesized that degree of control varies by two main sets of factors: the *demand* for workers' services in the labor market, and the amount of *resources* available to workers, which may be seen to enable greater power in obtaining job rewards (see also Kohn 1976). In general, elite scientists maintain high degrees of control over their work relative especially to communitarian scientists and a subset of pluralist scientists. In principle, there is also greater demand for the services of elites, whose performance records grant them greater mobility among employing institutions. Employed in resource-abundant institutions, elites also command a relatively greater share of organizational and professional resources, each of these sets of conditions thus heightening their opportunity for job satisfaction. In general, reversed conditions characterize the communitarian academic world. There is comparatively less demand for the services of communitarians, whose performance records render inter-institutional mobility less likely. Confined to institutions marked by relative scarcity of resources, communitarians are significantly restricted in access to resources that might otherwise grant opportunities for job satisfaction. Pluralists confront a mix of these conditions, although the conditions may be seen to sway toward one end or the other, depending on a department's location on the continuum.

The point at which the present findings appear to depart from Kalleberg's formulation consists in the element of time. *Despite a high degree of control over their work, elites grow disillusioned with work and its rewards at the end of their careers. Despite a relatively lower degree of control over their work, communitarians and pluralists grow more satisfied at the end of their careers once they have left, or near to the time of leaving, work* (proposition 4). These patterns also diverge from Kalleberg and Loscocco's (1983) subsequent theorizing with respect to job satisfaction and aging, wherein they postulated that job satisfaction increases with age. Removing the study of job satisfaction from its organizational environment and culture appears to black out significant variation in the conditions under which people experience work over a career. Moreover, by "bringing the organization back in," it is possible to see, in this instance, how some worlds of a vaunted profession may even begin to approximate forms of blue-collar work. By the end of their careers, and even well short of them, most communitarians and some pluralists hardly appear like dedicated, esteemed professionals.

What are the consequences of anomie to individuals and institutions? By its definition, clearly a marked consequence in individuals is the feeling of fragmentation, brought about by a "loss of faith in the value of one's own work" (Hagstrom 1964, 192). Hagstrom (1964; 1965), following Merton ([1957] 1968a), outlines five distinct types of adaptive behavior when recognition falls short of expectation. Scientists may engage in *retreatism*, withdrawing from creative work, renouncing both the goals and means of doing science. Retreatists may embrace alternative forms of reward in academic systems, such as in teaching or administration (see Glaser 1964b, especially p. 98–102). Scientists may engage in *ritualism*, continuing perfunctorily in research but not believing in its ends, a renunciation of the goals but not the means of doing science. Ritualists may continue in research to benefit salary growth, publishing a minimum to satisfy organizational expectations to merit annual salary raises. Scientists may engage in *innovation*, remaining committed to the achievement of recognition but attempting to win it by illicit ways, a renunciation of the means but not the goals of doing science. Innovators may plagiarize the work of others, attempting to gain credit for accomplishment where recognition is not due. Scientists may engage in *rebellion*, continuing to work, but by standards different from their main community of scientists, a renunciation of either or both the goals and means of doing science. Rebels may reject the standards by which their work is judged by the community of science and insist on their own, independent criteria to assess the importance of their work. Finally, scientists may engage in *conformity*, continuing to work according to both the institutionalized goals and means of science. Conformists continue to embrace research, believing in both the aims of science and the processes by which it is done.[1]

This study has observed varying manifestations of these adaptive strategies. Among communitarians, retreatism is most typical. Among pluralists, retreatism and ritualism are typical. Among elites, ritualism is most typical. Why might this be the case? Adaptation occurs within an organizational, not just a professional context. *Retreatism is most often enacted by communitarians and select pluralists because their universities offer a relative abundance of alternative rewards to research* (proposition 5a). One can count for something besides being an accomplished researcher. When the research career stalls, scientists turn to other legitimized outlets, afforded by their very own employers. *The reason one observes retreatism in greater combination with ritualism among pluralists is because the press for scientific achievement is greater in the pluralist world* (proposition 5b). While the pluralist world does offer alternative rewards, these are typically embraced with greater psychological and career costs. Hence pluralists are reluctant to completely give up on research; many plod along, attempting to maintain individual legitimacy. *Elites most typically embrace ritualism because anomie characteristically hits them late in the career; their productive habits are so well formed and so routine that they do not shake free from them, instead continuing to produce without the knowledge that they will not be as recognized as desired* (proposition 5c). All scientists—communitarian, pluralist, and elite—attempt to be conformists. But they deviate, at varying times, and adopt alternative ways of adapting to their unfolding careers. *Elites conform throughout the greatest portion of their careers* (proposition 5d).

Rebellion as an adaptive process appears most likely in environments that allow the greatest degrees of decoupling from the institution of science, namely the communitarian, and to some extent, the pluralist worlds (proposition 5e). Though not impossible, it is difficult for members of the elite world of science to decouple themselves from the institution of science, since elites socially control one another's careers in stringent fashion. To the extent rebellion was observed at all, it was among select communitarians and in even fewer pluralists who "became their own boss," declaring themselves to have rejected professional and organizational mandates to do science according to institutional norms. More often, though, such individuals were found to retreat; that is, they halted or significantly slowed their publication productivity. Rebellion, by contrast, entails a continuation of research productivity, albeit under the aegis of individually manufactured standards and goals.

Innovation appears rare, although notable cases outside the present sample of scientists heighten attention to its occurrence. Zuckerman notes that "basic evidence is lacking on the incidence, distribution, and effectiveness of various controls of deviant behavior in science" (Zuckerman 1988, 526).[2] The most egregious adaptive type of deviance from the scientific role, innovation likely entails costs too high for most scientists to adopt. It stands as the most extreme departure from scientists' socialized roles by including explicitly illicit and punishable conduct.

REFERENCE GROUPS AND SOCIAL CONTROL

A foundational contribution to the understanding of social organization in general, Durkheim's argument provided a framework for the subsequent development of a theory of reference groups in particular. Where Durkheim's anomie provides an answer as to *why* careers in academic science unfold the way they do, Robert Merton's formulation on the theory of reference groups helps to account for *how* this system of careers is sustained.

Reference groups are actual or imagined social categories of people with which individuals identify and make comparisons in guiding their beliefs, attitudes, and behaviors (Merton [1957] 1968a; [1957] 1968b; [1957] 1968c; [1957] 1968d). As Barbara Lawrence (2006) has explained, reference groups serve primary functions wherein individuals collect information from their frame of reference and in turn use such information to interpret their situations and guide their actions. Reference groups provide individuals with understanding about what otherwise would amount to ambiguous conditions and direct individual attitudes and behavior, in this case, attitudes and behavior about normative careers. In short, individuals collect information from others, sometimes near, sometimes far, sometimes living, sometimes dead, to interpret and act in their world of work as well as everyday life. Reference groups may be concrete and delimited in membership, such as a group of neighborhood spouses, a country club, a school class, or an academic department. Reference groups may also sometimes be imaginary or ambiguously defined, as in the case of deceased politicians, presidents or other cultural heroes, or members of professional pantheons, in which individuals postulate a perspective imputed to people (see Goode 1978).

> [S]cientists are idols oriented.... There is an implicit notion that there is a "real" scientist... doing the "real" scientific duty, the hard, backbreaking work of science... whose work is tied up with the main goals of science, its raison d'être, its conditions. Other functions are of lesser value; therefore the scientists who perform them are weaker and are less "real scientists." The model, then, is the ego ideal figure, who represents the ultimate position, and in fact, defines what... scientist[s] should do, how [they] should think, how [they] should act.... From this picture it is obvious that... scientist[s are] hard on themsel[ves]. [They have] built up a judgmental, critical superego which has a built-in, clearly marked scalar system, along which attitudes and kinds of performance are measured. When [they move] away and deviate from the pattern... [they become] maverick[s], or a person who has tossed aside the flaming torch. (Eiduson 1962, 167, 189–190)

As Shibutani put it: "Reference groups, then, arise through the internalization of norms; they constitute the structure of expectations imputed to some audience for whom one organizes his conduct.... One common usage of the concept is in the designation of that group which serves as the point of reference in making comparisons or contrasts, especially in forming judgments about one's self" (Shibutani 1955, 562, 565).

One has witnessed the manner in which scientists define success so as to make estimations and self-judgments of their achievement and professional-personal standing. One manner is internal, wherein scientists devise standards of self-worth using individually manufactured criteria, which can include projected perceptions of people such as family or friends. This manner of defining success and one's own achievement minimizes, and often altogether eliminates, the sanctioning role of professional peers in conferring and validating rewards and recognition.

Another manner of defining success is external, in which scientists arrive at estimations of their worth and achievement by invoking the standards of professional bodies. In their sanctioning role, these bodies utilize criteria of the institution of science that are used by groups and individuals to render judgments on others about the level and magnitude of achievement. The emphasis on recognition and Merton's explanation of it is recalled. Recognition serves as social testimony, conferred by the scientific community, that scientists have fulfilled the expectations of their professional role and thus the goals of science to extend certified knowledge (Merton 1973). All forms of recognition—publication, citation of published work, awards, honors, and so forth—are indexes of the scientific community's assessment of individuals' achievements as scientists. This manner of defining success and one's own achievement minimizes, and often altogether eliminates, the role of non-professional bodies, such as family and friends, in formulating estimations of achievement. The individual does not manufacture standards of success but turns to his or her profession for them.

One sees the manner by which scientists define success, over the phases of their careers, in the first panel of table 30. Also presented in the table, by the phases of the career, is the recognition scientists seek, the foci of their careers, and objects of their satisfaction.

As table 30 conveys, elites adopt external definitions of success throughout their careers. Pluralists waver between internal and external definitions of success at any given point in their careers or adopt one over the other. Not necessarily from the outset, but from early phases of their careers, communitarians utilize internal definitions of success.

The manners by which scientists define success may be interpreted as a selection of reference groups that scientists adopt to calculate their achievements (proposition 6). Elites turn to the great scientists, living and dead, those down the hall, across continents, or in the pantheon of science to guide their attitudes and beliefs about success as well their behaviors in their aspirations for greater achievement. While elites recalibrated their aspirations for professional achievement and grew more "realistic," their expectations for the future always remained substantial. By contrast, communitarians turned to themselves, as well as to family and friends, and above all, they turned away from science and scientists to render judgments on their achievement. Pluralists turn to themselves or to science to form judgments about their achievement, but at the end of their careers turn primarily to themselves.

Confirmatory evidence on these observations is offered by the recognition that scientists seek, and the foci of their careers in the various phases, as depicted in panels 2 and 3 of table 30. Throughout their careers, elites always seek at least substantial recognition from external bodies except when,

TABLE 30. Definitions of success, recognition sought, career foci, and scientists' objects of satisfaction, by career phases

SCIENTISTS' DEFINITIONS OF SUCCESS

Phases	Elites	Pluralists	Communitarians
Early to Mid	External	Uncertain	More Internal
Mid to Late	External	Internal or External	Internal
Late to Post	External	Primarily Internal	Internal

RECOGNITION SOUGHT BY SCIENTISTS

	Elites	Pluralists	Communitarians
Early	Great	Great	Great
Mid	Substantial	Average	Minimal
Mid	Great	Average	Minimal
Late	Substantial	Average or Minimal	None
Late	Great	Average	None
Post	Happenstance	Low; None	None

(continues)

at the end, they realize any additional recognition will occur as a result of happenstance. This is consistent with a career focus on research, which elites display over the phases of their careers.

Communitarians begin their careers seeking great recognition from external bodies, which by the end of their careers is reduced to a search for none at all. Correspondingly, they turn away from research and primarily embrace teaching.

Pluralists, too, alter the magnitude of recognition they seek, adapting from great to average to minimal quests for external recognition. Correspondingly, they, too, reduce the focus on research, but do maintain one alongside a more prominent teaching role compared to elites.

By the same token, the principal sources of scientists' professional satisfaction evolve, as depicted in panel 4 of table 30. Substantial evolution is seen among pluralists, and even more so among communitarians, who more firmly embrace teaching, mentoring and service activities of their roles. Over time, communitarians distance themselves from all profes-

TABLE 30. (continued)

CAREER FOCI OF SCIENTISTS

Phases	Elites	Pluralists	Communitarians
Early	Research	Research	Research
Mid	Research	Teaching/ Mentoring/ Research	Teaching/ Service/ Research
Mid	Research	Research & Teaching	Teaching
Late	Research	Teaching or Research	Teaching
Late	Research	Research & Teaching/ Teaching	Teaching
Post	Research	Light Research/ Not Professionally Active	Not Professionally Active

SCIENTISTS' OBJECTS OF SATISFACTION

Early to Mid	Research	Teaching/ Mentoring	Teaching/ Service
Mid to Late	Research; Administration	Teaching or Research	—
Late to Post	Research	Light research; Retirement	Retirement

sional roles to the point where, in mid- to late career, there is no real object of professional satisfaction and where, by the end, they most identify with retirement. The evidence points out shifting commitment: disavowal and distancing from research-related reference groups and identification with reference groups that affirm participation in more interpersonal and expressive roles, or roles outside of academia altogether.

This treatment of reference groups extends Alvin Gouldner's classic formulation of "locals" and "cosmopolitans" (Gouldner 1957–1958).[3] Locals are members of groups, such as organizations or communities, with high loyalty to their employing organization, low commitment to specialized skills, and internal reference groups. Cosmopolitans are members of groups who have little loyalty to their employing organization, high commitment to specialized skills, and predominantly external reference groups. The present work situates reference group behavior *temporally* in the scientific career. That is, the work has observed how reference group selection is apt to change in organizational contexts, particularly the communitarian and pluralist academic worlds. In Gouldner's terms: *some scientists realize both local and cosmopolitan identities over the course of a career; other scientists switch between them, gravitating toward one or the other in different career phases; and still other scientists embrace just one of these identities (proposition 7).* The existence of these career processes and identity transformations make for all but a static representation of reference-group behavior, and once again are suggestive of the organizational structures that shun or allow career-course continuity and change. Furthermore, whereas Gouldner attributes locals as having high loyalty to their employing organizations, the present work found contrasting evidence. *Many communitarians and some pluralists ended up bitter toward their employing organizations. And whereas they command and use specialized skills, and likely possess them well into their careers, the skills are not put to full use due to lack of opportunity and constraints that employing organizations place on professional careers (proposition 8).* Thus while the local-cosmopolitan distinction is useful in highlighting some major identity divisions among an organizational set (the array of colleges and universities, for example), additional career characteristics, such as those set forth in the tables, offer an elaboration about the temporal interplay among careers, identities, and institutions.[4]

For present concerns, the theoretic significance of reference groups lies in the groups' broader social function. By regulating, in greater or lesser ways, attitudes, beliefs, and behaviors, they function as mechanisms of social control (cf. Shibutani 1962). This points out the way in which scientists, elites in particular, sustain elevated expectations for professional achievement. *The gap between reality and desire among elites is always great because the reference group they utilize to form self-judgments is always far removed and itself an embodiment of greatness. The gap between reality and future wants among communitarians closes because the reference group they come to utilize to form self-judgments is of their own making (proposition 9).* By casting external definitions of success aside, communitarians avail themselves to friendlier judges—family, friends, and themselves—who invoke non-professional and hence less exacting standards to measure achievement.

In these respects, external reference groups composed of a profession's pantheon are more constraining on individual careers. In the case of elites, individual scientists experience comparatively less latitude in the activities that comprise their careers. Deviation is severely sanctioned. This is manifested in elites' singular research focus throughout the decades of a scientific career.

The ability of external reference groups to constrain individual behavior also establishes the conditions for comparatively continuous careers (proposition 10; cf. Hargens 1978). It is among elites that one observes the greatest career continuity. There are fewer fateful turning points, fewer reversals in outlook and orientation, less questioning of the goals of science, fewer complaints about work, colleagues, career, and institution. In still different terms, recalling Hughes, *if turning points are occasions to revise identities, then organizational environments that spell fewer turning points in careers will allow for greater continuity in individual identity, a pattern of which elites are most representative* (proposition 11). By contrast, organizational environments that spell greater turning points for careers create the ground for multiple change in individual identity, a pattern of which communitarians and select pluralists are most representative. External reference groups control careers and integrate individuals in institutional goals. A major irony is that this continuity and high integration punishes elites in the end. Comparatively stable in outlook and dedication, elites conclude their careers as the most discomfited.

By contrast, internal reference groups composed of family and friends and of oneself as one's own judge are more liberating for individual careers. In the case of communitarians and select pluralists, individual scientists experience comparatively great latitude in the activities that comprise their careers. Careers run a greater gamut. This is manifested in communitarians' and pluralists' lack of singular focus once beyond their early careers. In Hughes's terms, it is also manifested in the discontinuity of their professional self-identity over the course of a career.

The ability of these reference groups to liberate individual behavior establishes the conditions for comparatively discontinuous careers, for it is among communitarians and select pluralists that the greatest career discontinuity is observed. There are more fateful turning points, more reversals in outlook and orientation, greater questioning of the goals of science, more complaints about work, colleagues, careers, and institution. Internal reference groups decouple individuals from institutional goals. Another major irony is that this discontinuity and weak integration frees communitarians and select pluralists in the end. Comparatively unstable in career outlook and dedication throughout almost all of their careers, communitarians and select pluralists conclude them as the most at peace.[5]

SELECTION OF REFERENCE GROUPS

This view of the system of scientific careers—and academic careers generally—prompts a question of reference-group selection. Merton and Zuckerman discussed the general problem of how reference groups are selected by the scientists and scientific groups who invoke them (Merton and Zuckerman 1973). The answer to which they turn lies in the reward system of science: scientists estimate gains and losses in continuing or cutting back in publication productivity, and they may act upon values toward research work that change in light of allocations of recognition. The highly productive may remain productive because the promise of greater recognition or, conversely, the stigma of slowing or stopping remains strong. The less productive may turn to activities other than research because the promise of still greater recognition is more elusive and the stigma of slowing or stopping less severe. As Stephen Cole has stated: "As [scientists] continue to publish, some find their work rewarded and go on to publish more . . . those who are not rewarded are less likely to continue publishing. Thus, as a cohort of scientists advance in age the number of prolific publishers is likely to decline. Most people will not continue an activity as arduous as scientific research unless they are rewarded for it" (Cole 1979, 969). The present findings are consistent with this view but point to a further source for theoretic understanding.

The evidence on hand suggests that *reference-group selection depends not exclusively on the operation of a professional reward system (i.e., the reward system of science) but also on organizational reward systems, those situated in the departments and universities that employ scientists* (proposition 12). In strict terms, Merton and Zuckerman offer a *professional* interpretation of reference-group selection, geared to the community of scientists. Here, reference-group selection is *organizationally* situated, geared to the institutions in which scientists work, while also responsive to the professional workings of rewards.

It is not that any scientist is at home in any university. Different work contexts are not merely strata formed in response to a system of reward—they have different purposes, many of which are unallied with research or the operation of the scientific reward system. Following Hughes (1958), individuals and institutions do their mutual creating of one another. What would be created if half of the sociology faculty from the University of California–Berkeley were transplanted to the University of Toledo, and vice versa? Would the Berkeley faculty in Toledo take to a heavier teaching load, with little or no discretion in the time of day they taught? Would they find it okay to teach large course sections with few if any graduate assistants? Would they enjoy the comparative lack of resources for, or press to conduct, research? Would they mind not teaching graduate students or not chairing doctoral committees? Could they handle the lack of attention and prestige? And for the Toledo faculty now at Berkeley: would they take to pressure to publish? Would they be agreeable to focusing on graduate students and graduate education, even if it meant sometimes coming at the expense of undergraduates? Would they mind the press to get external grants, or to be "public" and visible in their scholarship?

By even early points in their careers, scientists (like perhaps all academics) are embedded within an organizational structure and culture that constrains and conditions their careers—as well as individual goals, outlooks, attitudes, and behaviors—in ways different from one type of university to another. To this end, reference-group selection is organizationally situated and determined. Elite, pluralist, and communitarian universities possess different missions. Through individual careers, the institutional mission in the elite world is expressed as "excel in research." In the pluralist world, it is "excel at something." In the communitarian world, "excel first at teaching, then turn to research if time allows." The differences in these organizations stipulate the reference groups to which scientists turn to shape their behavior in and attitudes about their roles. Among elites, a high-achieving external reference group is consistent with the organization's needs and purpose as a leader on the higher education frontier. Among communitarians, a distancing from high-achieving external reference groups and embrace of an internal one is consistent with the organization's needs and purpose to teach a regional supply of undergraduates and a still more modest regional supply of graduate students. Among pluralists, a mix of external and internal reference groups is consistent with the organization's needs and purpose to be many things to many people and many constituencies: an organization of research, teaching, extension, and public service, the modern multi-university. On these terms, performance strata are constituted not only by the operation of a professional reward system, but also organizational culture and structure: especially in communitarian and pluralists institutions, academics can *learn*, from the time they enter their institutions, and especially once past tenure, how to be acceptably unproductive.

It is its blend of the elite and communitarian extremes that gives definition to the pluralist world of academia, situated in the middle of the departmental and institutional continuum discussed in chapter 1. Singularity of form—that is, something strictly elite or strictly communitarian—is arguably more straightforward than a mix—a plurality. Pluralist universities, from the University of Georgia to the University of Arizona, are the largest academic organizations in the world, with more faculty and students, both graduate and undergraduate, than any other type of university. To this end, many different and often disparate demands are placed upon them. In attempting to satisfy numerous purposes and numerous constituencies, they employ faculty members who, over the courses of their careers, gravitate in one direction (teaching, for instance) over another (research). They sometimes change those directions at various points, especially after early career, when academic tenure has been secured or all promotions have been obtained (leaving a research career for full-time administration, for example, or deemphasizing a research career in favor of undergraduate teaching).

The fact of the matter is that pluralist institutions require these roles: they profess a need for not just singularly focused researchers, though they need them to be sure. They also profess a need for award-winning teachers to satisfy parents, politicians, and legislatures, and administrator-servants who abandon their research and teaching to manage large, multifaceted organizations. In different universities, many of these pluralists would be fish out of water; life on the Bay moved to the industrial shores of Lake Erie and vice versa, a foreign exchange. Thus, various reference groups are available to guide pluralist faculty members—their attitudes, beliefs, and behaviors—in their careers. The multiple reference groups are legitimate in this world, owing to the world's multiple missions. In this light, *plurality of legitimized choice can make it difficult to fail* (proposition 13a). It is for these reasons that pluralists sometimes regard their departments and institutions as a happy medium. It is also for these reasons that one finds careers structured by scientists' employing organizations, and not only the reward system of science.

Further, reference groups are not the only bodies that control careers in their organizational contexts. Reward structures of organizations, apart from an encompassing reward system of science, do as well. Yet it appears as though, more often than not, reward structures of organizations parallel the form and function of the reference groups that those organizations specify for individuals in the enactment of their careers. Thus, in elite departments and universities, one is apt to find reward structures greatly, if not exclusively, favoring research. In communitarian departments and universities, one is apt to find reward structures that pay serious attention to teaching, while also rewarding research, but not if it comes at the expense of teaching.

In pluralist departments and universities, one is likely to find a mixture of these structures, as if to form separate "career tracks," again responsive to the multiple needs and purposes of the pluralist institution. Members of pluralist departments know quite clearly who are the dedicated and successful researchers and who are the dedicated and successful teachers. On occasion, they can be one and the same. But, having worked around them for decades, most individuals can detect a predilection in their colleagues. The proportion of each will vary depending where on the continuum a department falls (see figure 1). When closer to the elite end, more are dedicated researchers. When closer to the communitarian end, more are dedicated teachers. The larger point is that individuals can derive institutional rewards in the pluralist world from either type of career emphasis, learning the ways of research productivity on the one hand or the ways of acceptable unproductivity on the other. This explains why one hears research-oriented pluralists remark (often with some astonishment) at how teaching-oriented pluralists improve their salaries through teaching performance and awards. It also accounts for the friction and animosity that can develop between camps. Research-oriented professors desire structures that favor research. They typically see teaching as an appendage to their research. Teaching-oriented professors do not want reward structures favoring research to get too extreme—they clamor that the university has an obligation to the citizens of the state. Each seeks to preserve and defend their identity in the face of apparent threat.

The opportunity for reward in research will grow as a department is located toward the elite end of the spectrum; the opportunity for reward in teaching will grow as a department is located toward the communitarian end of the spectrum. But whether located toward either end or in the middle of the continuum, the pluralist department and institution will sanction, albeit in varying degrees, varieties of careers, since varieties of careers are alleged to be necessary for the organization's survival (proposition 13b).

REJECTION OF REFERENCE GROUPS

While organizationally suited reference groups control careers in their respective settings, not each and every scientist in a given setting conforms to the organizationally suited reference group. In any professional career, just as in any group, there is deviation (proposition 14). Remarking on criminals and their breach of the law, Durkheim ([1895] 1982) saw functionality in the deviant role: groups need deviants to remind them of customary order and the normal course of life, for what is normal in the absence of any demonstrated contrast? This is why torture of civilians by the military is both shocking and functional.

It is shocking because it is a breach of ethical conduct within the military profession and functional because it reminds the military of how to properly perform its role. Likewise, in civilian society, murder is alarming, but simultaneously calls attention to the sanctity of human life. Deviation in the scientific career is thus expected and necessary to science.

Among elites, some scientists markedly slow down or stop their research. Among communitarians, some scientists steer clear of teaching to embrace and sustain research. These types are deviants in their respective worlds. Notable among the deviants, in all cases, is that they remained deviant from the time of the foundational to that of the longitudinal study. None reverted to the more typical path followed by other scientists in their respective academic worlds. *The finding of sustained deviance further suggests the stricture with which the academic profession structures careers: once a deviant, always a deviant* (proposition 15).

This may be explained by prior theory. Cumulative advantage stimulates further productivity in the communitarian who receives disproportionate recognition compared to local colleagues (Dannefer 1987). Early advantages spell further advantages as the career develops an ever firmer research course. Cumulative disadvantage mutes further productivity in the elite who stops or slows markedly in research. It is difficult to fund, honor, or otherwise reward a scientist, even if he or she is a member of the elite academic world, after an unconcealable dry period. As Merton and Zuckerman (1973) noted, allocation of reward under these conditions may be especially difficult with elites, since the expectation for performance is as substantial as the disappointment created by the dashed expectations.

Notable as well as expected is that deviants know they are deviant. It is difficult to violate norms without knowing they have been violated. True as this may be, the occurrence provides phenomenological evidence of deviation in the scientific career, when and how it occurs as well as its construal by self and others. Like all deviation, it calls attention not exclusively to the pathological but also to "the normal order of things." Hence one can on rare occasion encounter the elite physicist who shuns research in mid- and into late career in favor of a communitarian-like embrace of undergraduate teaching:

> This is a research university. Despite all the lip service, undergraduate education is not considered important. I don't have the respect of my peers. Teaching is not respected. Teachers are people who can't do research. (12L)

By the same token, one can on rare occasion encounter the communitarian physicist who shuns teaching as far as possible from early career onward, to the present point in late career, in favor of elite-like research:

> Without grants, you end up with a nine-month salary, and you have to ask the department chair if you can teach a class during the summer, which takes up your summer. I've never taught in the summer. I've always been able to fund myself during the summer for the past thirty years. Every month of every summer I've been here, I've been paid. Not everybody here can say that. There are a lot of people who don't have money during the summer, so they have to teach a course. (25L)

The elite scientist rejects an external in favor of an internal reference group, the communitarian substitutes an internal for an external reference group. Change of reference group entails change of career expectations. The elite scientist finds satisfaction in the teaching role, the communitarian challenge in the role of researcher. Moreover, just as one observes substitution of reference groups and the types of expectations that accompany them, one observes anomie and its presence (or lack thereof) in the scientists. The elite scientist finds a communitarian-like comfort typical of the latter-most career phases among communitarians (while also well aware of his deviance and the stigma associated with it).

> I'm more outwardly directed. I'm more social and less bound up in a small research community where I don't have much profile outside of the research community. . . . I'm happier now. . . . I've been as successful as my talents would justify. (12L)

By contrast, the communitarian-deviant develops an elite-like desire for more achievement.

> My job has become totally consuming. . . . I'm working harder. . . . I work every weekend. . . . Enough is not enough. [I want] more publications. More grants. Tangible things. (25L)

One hears the echo of Durkheim: "Inextinguishable thirst is constantly renewed torture" (Durkheim [1897] 1951, 247).

SOCIAL CONTROL OF THE LIFE COURSE

At any given time in their careers, the shifting patterned perspectives of scientists may be viewed by the way in which their professional life courses are socially controlled via organizational reference groups. Reexamining the overall modal career patterns of scientists over time across the academic worlds of science, as presented in the first panel of table 29, one observes comparative stability among elites (except at the end of their careers) because of consistency of reference-group selection. This consistency is enabled by a relatively unitary organizational mission.

Greater career discontinuity is observed among pluralists and communitarians because of a multiplicity of available reference groups, enabled by organizations with multiple missions. As Howard Becker has noted: "The process of situational adjustment suggests an explanation of change; the process of commitment suggests an explanation of stability" (Becker 1964, 40). It is possible for pluralists, for instance, to reverse their career outlooks and orientations as they proceed from early- into mid-career phases because an organizational mandate to keep achieving in scientific research competes with other career possibilities suited to the organization. With tenure, pluralists encounter other possibilities for their time besides that which heretofore guided their careers. "[T]he individual turns himself into the kind of person the situation demands," and which is afforded by the social structure of the organization of which that individual is a part (Becker 1964, 44). Pluralists continue to face these competing possibilities throughout their careers, as when in mid- to late-career transitions, they either continue on the research courses they have firmly established or attempt to regenerate research-focused careers following deceleration in research and reevaluation of how to envision their future.

Among communitarians, by early- to mid-career transitions, with twenty or more years of an academic career ahead of them, the adaptation to local custom has already achieved deep roots. The modal pattern is stasis. Communitarians display little behavior of professional advancement in scientific research, consistent with an internal reference group. This pattern develops further such that in middle to late transitions the overall modal pattern of careers is that of demise.

While analytically and operationally distinct from the profession's system of scientific reward, these organizational dynamics are of course not wholly independent of the reward system of science, itself a mechanism by which careers are socially controlled. Processes of cumulative advantage are underway among elites and select pluralists and even more select communitarians who have developed track records as successful researchers. Conversely, processes of cumulative disadvantage are underway among communitarians and select pluralists who are less successful in research, become less productive, and whose progressive unproductivity is reinforced by this adverse feedback process.

The processes of cumulative advantage and disadvantage are expressed not only in the publication productivity patterns observed as scientists age (tables 20, 23, and 26) but also in their changing subjective views toward the reward system of science. Scientists' perceptions of the scientific reward system, coupled with their orientations to work and work intensity, are presented by career phases in table 31.

Over the courses of their careers, elites especially view the reward system of science as fair. Only by late and post-career do elites find fault with the system in great measure, owing to the anomie they experience at these points in time. By contrast, communitarians view the system as unfair throughout nearly all of their careers, their earliest years in science perhaps possessing their most favorable views toward the reward system, prior to its full engagement and operation with felt costs. Pluralists waver considerably more. The system is viewed as unfair as pluralists undergo a reversal in the modal patterns in early- to mid-career phases. They are split in mid- to late-career phases as here, too, their modal careers are divided between continuation of scientific research on the one hand and, on the other, attempted regeneration in their research. By the end, pluralists regard the system as fair overall, but have withdrawn from it in their overall career pattern.

The patterns of scientists' perceptions of the scientific reward system are in turn associated with distinct patterns in the orientations they develop toward their work and in the self-judgments they make about how hard they work in the three clustered phases of their careers. A perception of the reward system as fair is associated with an enduring moral commitment to science and scientific work, as evident among elites (illustrated in panel 2 of table 31). It is also associated with an intensifying, or at least consistently strong, work intensity, also apparent among elites (illustrated in panel 3 of table 31).

By contrast, a perception of the reward system as unfair is associated with a crippled moral orientation to science, in turn supplanted by a utilitarian outlook on work, characteristic of communitarians, in which a profession evolves into "a job." Correspondingly, communitarians state they work "less hard" on science.

TABLE 31. Scientists' perceptions of the reward system of science, orientation to work, and work intensity, by career phases

SCIENTISTS' PERCEPTIONS OF THE REWARD SYSTEM OF SCIENCE

Phases	Elites	Pluralists	Communitarians
Early to Mid	Fair	Unfair	Unfair
Mid to Late	Fair	Unfair; Fair	Unfair
Late to Post	Faulty	Fair	Unfair

SCIENTISTS' ORIENTATION TO WORK

	Elites	Pluralists	Communitarians
Early	Moral	Moral	Moral
Mid	Moral	Moral or Utilitarian	Utilitarian
Mid	Moral	Moral or Utilitarian	Utilitarian
Late	Moral	Moral or Utilitarian	Utilitarian
Late	Moral	Moral or Utilitarian	Utilitarian
Post	Intellectual	Utilitarian	Removed

SCIENTISTS' WORK INTENSITY

	Elites	Pluralists	Communitarians
Early to Mid	As Hard; Harder	Less Hard	Less Hard
Mid to Late	As Hard; Harder	Less Hard; As Hard	Less Hard
Late to Post	As Hard; Less Hard	Less Hard	Less Hard

Among pluralists, the mix of perceptions toward the reward system of science is associated with a corresponding mix of orientations at the onset of mid-career. Dedicated and successful researchers, presumably beneficiaries of accumulating advantages, remain morally oriented to science. Less dedicated and less successful researchers, presumably subject to accumulating disadvantages, develop a utilitarian orientation to their work. In general, pluralists state that they progressively work less hard, suggesting that those remaining engaged in research do so but at a perceived slower, or simply perhaps a more peaceable, pace, and that those disengaging from research also experience a lessening of work intensity. *The logic that emerges from the data strongly suggests that scientists develop the observed behavioral responses to their careers as their work, over time, is variously and disparately recognized by reward systems* (proposition 16). Reward systems, professional and organizational, thus operate as chief causal mechanisms underlying the observed career patterns of scientists.

Also noteworthy is the extent of discontent with the reward system of science: *it is widespread, found across all the worlds of science and in all the cohorts of scientists* (proposition 17), notwithstanding the pockets of relative content found in early and mid-career among elites and in mid- and late career among select pluralists. The data convey that grievance with the scientific reward system is a near ubiquity. Scientists' ambivalence with the reward system of science has three main expressions: scientists' beliefs that recognition for accomplishment is delayed; scientists' beliefs that the amount of recognition does not correspond to the level of achievement it rewards; and scientists' beliefs that recognition is at times not granted at all for genuine accomplishment. In all three expressions of grievance, concern lies with the sufficiency of recognition. *Widespread concern about the deficit of recognition provides further indication of the centrality that recognition plays in the construction, meaning, and evaluation of an academic career* (proposition 18a).

In her study of Nobel laureates, Zuckerman (1977) found few scientists who believed recognition of their work had been delayed, a finding that perhaps owes itself in part to the stratum of scientists studied, and one which is also consistent with the present findings. The most rewarded scientists possess the most (though not uniformly) positive views toward the reward system of science throughout the greatest portion of their careers. Moving out from the elite, however, the present data suggest that this attitude splinters greatly. The profession is both unequal in who receives recognition and in the attitudes of its members toward the system that allocates it.

This is, of course, not evidence that the system of reward is objectively unfair or wholly malfunctional; several earlier studies have presented evidence to the contrary (e.g., Cole 1970; Cole and Cole 1967; Zuckerman 1977). Rather *the data convey a ubiquity of a viewpoint, suggestive of the character of the profession, that scientists regard their work as more significant than conveyed by the institutional recognition their work receives. Such a viewpoint establishes another ground for anomie, since individual beliefs about rewards deserved exceed the reality of rewards granted* (proposition 18b).

The data further convey that this viewpoint, while dissipating in variant degrees across organizational contexts, evolves into correspondingly vari-

TABLE 32. Scientists' perceptions of peak career, professional aspirations, work/family focus, career progress, and attributions of place, by career phases

SCIENTISTS' PERCEPTIONS OF PEAK CAREER

Phases	Elites	Pluralists	Communitarians
Early to Mid	Present; Postdoc	Early Career	Present; Postdoc
Mid to Late	Present	Early Career; Present	Early Career
Late to Post	Early Career	Present; Mid-Career	Present; Early Career

SCIENTISTS' PROFESSIONAL ASPIRATIONS

Early	Intensify	Rescaled	Diminish
Mid	Intensify	Diminish	Subside
Mid	Intensify	Diminish	Subside/Extinguish
Late	Recalibrated	Subside or Remain Steady	Nonexistent
Late	Remain Steady	Subside	Not Present
Post	Tempered	Subside	Not Present

SCIENTISTS' WORK/FAMILY FOCUS

Early	Work	Work	Work
Mid	Work	Family & Work	Family & Work
Mid	Work	Family & Work	Family & Work
Late	Work	Family & Work	Family & Leisure
Late	Work	Family & Work	Family & Leisure
Post	Work & Leisure	Leisure & Work;	Leisure Leisure

(continues)

ant residues in scientists' work perspectives. Such a common, remarkably negative, and long-lasting sentiment would appear to be a structural weakness of the institution of science, since on the one hand the reward system attempts to motivate effort but on the other hand ineffectively handles the legions of those whose efforts apparently fall short. One is able to see, in organizational context, the new and changed attitudes that the profession

TABLE 32. (continued)

SCIENTISTS' CAREER PROGRESS

Phases	Elites	Pluralists	Communitarians
Early to Mid	As Expected; Better than expected	Not as Expected	Not as Expected
Mid to Late	As Expected; Better than Expected	Not as Expected; Better than Expected	Not as Expected
Late to Post	Not as Expected	As Expected	As Expected

SCIENTISTS' ATTRIBUTIONS OF PLACE

Early	"Burden"	"Happy Medium"	"Stymieing"
Mid	"The Best"	"Den of Confusion"	"A Job"
Mid	"Haven"	"Happy Medium"	"Stymieing"
Late	"Haven"	"Place to Succumb or Triumph"	"Hapless Bureaucracy"
Late	"Haven"	"Happy Medium"	"Stymieing"
Post	"Haven"	"Place of Work"	"Place Departed"

instills in the majority of scientists who do not achieve great fame, yet their distaste for the reward system of science can always be detected. The institution of science, in conjunction with academic organizations, functions to make such distaste more latent than manifest, but it appears that merely dyadic interaction, such as through an interview about the scientific career, can make this attitude and behavioral response more manifest than latent.

Given that one observes in this work generally widespread ambivalence among scientists toward the reward system of science, why do different groups of scientists report different "peak points" in their careers? If the vast majority of scientists develop discontent with the reward system of science, one might expect them to identify a more uniform "peak" in their careers, or that the "peak points" would be random, an essentially individual feeling unshaped or formulated by social structure. Yet one observes distinct patterns in scientists' perceptions of their peak careers, depicted in table 32, together with their professional aspirations, work/family focus, career progress, and how they view their universities, by career phases.

The scientists most likely to identify a prior point in time as their career peak are pluralists and communitarians, whereas for elites the peak career tends to move as scientists work and be located in the present. In general, pluralists and communitarians are most apt to turn to the past and elites to the present in locating themselves "at their best."

It is important to note that these patterns and the question that prompts them—"When do scientists perceive themselves at their peak?"—are related, but very much distinct from past work on the more general topic of age and work performance (Zuckerman 1988). In science, as in other institutional realms, a common belief is that individuals do their best work when they are young. It may also be believed that, in the case of science, individuals are most productive in publication in their younger years. These notions were lent credence by the work of the psychologist Harvey Lehman (1953; see also Stephan and Levin 1992), who thought that there was a relationship between being young, creativity, and achievement. Despite the strength of these beliefs, now diffused throughout culture, empirical evidence does not support them (Bayer and Dutton 1977; Reskin 1979). Moreover, Lehman's work has since been found to be methodologically flawed (Cole 1979).

Instead, research has found that the relationship between age and scientific performance is modestly curvilinear; publication productivity is generally seen to increase modestly as scientists enter middle age and then decrease modestly as they advance further in age (Cole 1979). As a case in point, Zuckerman recorded that physicists were on average 36.1 years of age when they did the research that resulted in a Nobel Prize. Chemists were 38.8, biological scientists, 41.1 years of age on average (Zuckerman 1977, 166). Organizational contexts of academic employment affect productivity; a concentration of highly productive colleagues creates an environment to stimulate high levels of research publication (Allison and Long 1990; Braxton 1983; Crane 1965; Long and McGinnis 1981; Pelz and Andrews 1966; Reskin 1977). Further, processes of cumulative advantage and disadvantage differentiate scientists as they age (Cole and Cole 1967; 1973). As Allison and Stewart (1974, 596) observed: "Because of feedback through recognition and resources, highly productive scientists maintain or increase their productivity, while scientists who produce very little produce even less later on. A major implication of accumulative advantage is that the distribution of productivity becomes increasingly unequal as a cohort of scientists ages," a pattern that was observed in the publication productivity of the present sample of scientists, as indicated in tables 20, 23, and 26. Some publication patterns vary by academic field (Bayer and Dutton 1977; Wanner, Lewis, and

Gregorio 1981) and by gender (Cole 1979; Cole and Zuckerman 1987; Fox 1985; 2005; National Research Council 2001; Reskin 1978b; Sonnert and Holton 1995; Xie and Shauman 2003; Zuckerman and Cole 1975; Zuckerman, Cole, and Bruer 1991). Overall, however, the evidence does not point to a strong relationship between youth and doing one's best work.

How the reward system of science and how organizational settings of academic work influence productivity patterns have already been observed. In the case of the reward system of science, when early work is recognized, scientists are apt to continue to be productive, whereas scientists whose early work is unrecognized are apt to become less productive, each respective process of advantage and disadvantage reinforced over time (Allison, Long, and Krauze 1982; Cole and Cole 1967; 1973; Zuckerman 1970; 1977). In the case of organizational settings of academic work, available constellations of roles (for example, greater research over teaching, as among elites; greater teaching over research, as among communitarians; research intensity and then retreat, as among select pluralists) together with varying university missions for performance, foster or impede a research career.

Zuckerman (1988) explained why it might be tempting to believe that there is a relationship between age and achievement. As so often is the case for numerous purposes, the pantheon of science is invoked, enabling one to turn to the Newton at twenty-four for the invention of calculus, to the Einstein at twenty-six for the elaboration of relativity, to the Darwin at twenty-nine for the theory of natural selection, and so on (Zuckerman 1988, 533 – 534; also discussed in Zuckerman and Merton 1973 and Cole 1979). Rarely are epoch-defining events, or epoch-making individuals, generalizable to wider populations who follow in their long wakes, despite the inspiration that such events and individuals provide for those who follow. Such is true regarding the relationship between age and scientific achievement.

What, then, accounts for the patterns in which scientists perceive themselves at their peak careers? To be sure, one source could be the pantheon itself, stirring the imagery of age and achievement for those who have indeed followed in the wake of heroes. But one would expect elite scientists to identify their early careers as their peak just as many pluralists and communitarians do. For a more complete answer, I turn to the sociologist Erving Goffman.

In a classic essay, Goffman (1952) put forward the idea of "cooling out the mark." Using the metaphor of confidence games, Goffman explained that much of society operates on individuals' socially produced expectations, which must be reconciled within a social structure that constrains opportunity and ability for individuals to realize those expectations. The mark is the one whose expectations have been raised; the cooler "takes" but must ultimately console the mark, so as not to leave the mark shattered and functionless or, in the con, dubious of the exchange in which the mark (unknowingly) has been taken. Typically, the cooler makes a "substitution" on behalf of the mark in order to pull off the con, offering a consolation prize in place of that which the mark originally desired. In those instances when cooling-out proceeds smoothly, the substitution is made to appear in the mark's eyes as a prize just as good if not better than what the mark originally expected. Implicit in Goffman's argument is a virtue underlying the cooling-out process: it leaves individuals in greater alignment with reality.

Cooling-out techniques and processes are observable in all social institutions and in numerous social situations: the child in the toy store whose expectations for a large toy going in are reconciled by the tactful parent into a less expensive, but much "nicer" toy coming out; the graduate student whose expectations for a job suited to his training reconciled by a mentor for a job less suitable, but possessing other, now more important advantages, such as location and cost of living; the teenager who is asked by his girlfriend to be just a friend, since there is "so much more" to friendship; the couple expecting years of romance at the outset of their marriage only to be told by society, operating occupationally through the roles of counselor and psychotherapist, that a divorce would allow them "to be happier" (for selected systematic treatments, see Ball 1976; Faulkner 1974, 1985; Goldner 1965; Goldner and Ritti 1967). Little is different in the case of the occupational arena:

> Personnel come to define their career line in terms of a sequence of legitimate expectations and to base their self-conceptions on the assumption that in due course they will be what the institution allows persons to become.... A mark's participation in a play, and his investment in it, clearly commit him in his own eyes to the proposition that he is a smart man. The process by which he comes to believe that he cannot lose is also the process by which he drops the defenses and compensations that previously protected him from defeats. When the blowoff comes, the mark finds that he has no defense for not being a shrewd man. He has defined himself as a shrewd man and must face the fact that he is only another easy mark. He has defined himself as possessing a certain set of qualities and then proven to himself that he is miserably lacking in them. This is a process of self-destruction of the

self. It is no wonder that the mark needs to be cooled out and that it is good business policy for one of the operators to stay with the mark in order to talk him into a point of view from which it is possible to accept a loss.... The mark must therefore be supplied with a new set of apologies for himself, a new framework in which to see himself and judge himself. A process of redefining the self along defensible lines must be instigated and carried along; since the mark himself is frequently in too weakened a condition to do this, the cooler must initially do it for him. (Goffman 1952, 452, 455–456)

Scientists cool out, but do so at different times, at different rates, and with different consequences, depending on the type of organizational environment in which the process occurs (proposition 19a). While relatively mild forms of cooling-out are apparent among all scientists (as perhaps among all individuals), the evidence suggests that communitarians cool out in major ways the earliest in their careers, followed by pluralists, then elites. This pattern corresponds to the opportunities available across the academic settings for professional development, that is, the cultivation of a research career consonant with the institutional goals of science. The fewest such opportunities are found in communitarian settings, more are found in pluralist settings, and are abundant in elite settings.

In these terms, when scientists look back, they look at the points at which they still appeared in their eyes to have the greatest promise in realizing their goals. For many scientists, the early career consists of a time full of professional promise, and short of the point at which their aspirations and corresponding self-conceptions have, in Goffman's words, been blown off. For elites, however, the environment, through its reference groups and accumulating rewards, enables scientists to sustain exalted beliefs about their future. This is why they routinely perceive themselves to be at their best in the present. It is only at the end of their careers, when the finality of their not joining the reference groups that guided their careers settles in, that they alter their temporal frames and locate their best years in the past. And they do so at a particular point in the past, the early career, because it is then, more so than at any other time, when the future—and their anticipated place in it—knew no limits. Various pluralists and communitarians identify the present as their peak in post-career because they have been freed from expectation altogether.

From the standpoint of the reward system of science and its control of careers, these cooling-out patterns may well overlap with inflection points at which cumulating advantages and disadvantages begin to take hold. The corresponding paths of the scientists' professional aspirations, depicted in table 32, offer support for the point. For elites, professional aspirations intensify, adjust, and remain steady, mirroring a pattern of early advantages that spirals over time. Much greater discontinuity is found among pluralists and still greater change among communitarians, whose professional aspirations subside sharply even as early as the onset of mid-career. Communitarians let go of virtually all professional aspiration in the passage from mid- to late career, points at which disadvantages spiral and make the prospect of further scientific work probabilistically susceptible to the same neglect by the scientific community. Scientific research is seen as too arduous to undertake in light of the few anticipated rewards.

The "substitutions" that occur in the cooling-out process can take varieties of forms. One observed (in table 30) the teaching and service roles, commitments, and orientations that scientists increasingly adopt, particularly in the communitarian and pluralist worlds of science, at the end of early-career phases once jobs are secured through tenure. One heard of administrative roles assumed and new emphases placed on mentoring and other interpersonal aspects of roles adopted by scientists as their careers unfold.

Family and leisure, depicted in the third panel of table 32, are other forms of substitutions. As rewards from work decline, one's identity may be progressively staked in varied terrain. The pattern one observes is that as scientists cool out, they increasingly discuss family and leisure as relevant aspects of themselves to discuss in an interview about work and career. This further accounts for why one observed scientists discussing the importance of family and friends in their self-evaluations about work. They are internal reference groups that take away, or at least soften, the pain of what otherwise would appear in scientists' eyes to be failure in the scientific role. The phenomenon of turning to family and leisure when expectations of the rewards of work fall short appears to be common (see Faulkner 1974). It does, however, constitute a marked change in the self and calls forth an entirely different conception of roles and the meaning of one's performance in them. Scientists' diachronic accounts suggest that this change takes years, and often decades, to accomplish, and then only with a residue of what was left behind. The fourth panel of table 32 illustrates the ways in which scientists accounted for the progress of their careers. For communitarians and a subset of pluralists, the career does not go as expected, illustrating the lengthiness of the cooling-out process. Only in very late and post-career is it seen to have gone as expected — the cooling-out process complete. Among elites, a substantial component of the cooling-out process *begins* in the latter-most phases of the career when, with the greatest finality, expectations exceed the opportunity to satisfy them.

It is little coincidence that the patterns of cooling-out across the settings of science are associated with distinct attitudes that scientists develop toward their employers. The fifth and final panel of table 32 captures the ways in which scientists regard the universities in which they have attempted to work. Only elites are able to maintain, for the most part, positive sentiments toward their universities, which they see as having facilitated their efforts. Among pluralists, and again especially among communitarians, assessments of places of work grow more negative, accusatory, and cynical. It is clear that for many the cooling-out process leaves individuals maladjusted in their attitudes toward their work environments.

In Goffman's analysis, cooling-out appeared to possess a fairly fixed and relatively truncated temporal duration, illustrated in his use of job-training personnel as examples, wherein more people are turned away for company positions than can be hired. The analysis also conveys a successful, nearly seamless performance of the cooling-out process, an endemic feature of modern societies wherein one finds an endemic of individual expectation for the future (McClelland 1961; Merton [1957] 1968b, especially p. 220–224). It is so ubiquitous and performed so frequently, and thus institutionalized in social structure, that the process appears routine.

The occurrence of the process in academic careers, however, evinces striking contrasts. Here, the process is not at all brief and seamless. Indeed, the longitudinal data convey that the cooling-out of professional expectation in science typically extends over decades, and does so by customarily leaving a bitterness in the mouths and minds of the many subjected to it. Not quite the process Goffman had in mind. Why does it occur this way?

The time-consuming nature of cooling-out in science, and in academia more broadly, arises for specific reasons. Goffman's discussion highlighted the roles of all involved in the process, including the cooler, in bringing about a successful resolution of the mark's identity. But *in the case of science, much if not all of the cooling-out process takes place independently. The mark and cooler are one and the same* (proposition 19b). It is true that some scientists may turn to trusted colleagues for support and encouragement in times of great stress, but most can never confess the depth of their grief or the actual extent of their expectations. This may be interpreted as a structural flaw of the cooling-out process in professions. Why does one come upon this "flaw" in professional occupations?

Work relations in science, as in many professions, tend to be predicated on universalistic, professional grounds rather than particularistic, expressive ones. While recognition is necessary to the operation of science, scientific norms suppress overtly public desires for it. This is conveyed in an overriding tension of the scientific role explained by Merton: "New scientific knowledge should be greatly esteemed by knowledgeable peers," but, "The scientist should work without regard for the esteem of others." "Contributing to the substitution of sentiment for analysis is the often painful contrast between the actual behavior of scientists and the behavior ideally prescribed for them" (Merton 1976, 33, 40). Practitioners are not supposed to discuss their griefs; it is contrary to performance norms. Hence much of the trauma is tied up in individuals. Short of an effective cooler, marks are never fully cooled out. This helps to explain why pluralist and especially communitarian scientists progressively regard their institutions in negative, cynical, and then ultimately, in detached terms. *Structurally, it is difficult for members of a profession to discuss failure, and its many members are thus left partly paralyzed by it. The ineffectiveness of the cooling-out process in academia disallows the potential of completely "sending failure away," of effectively funneling it outside the social organization of work* (proposition 19c). Under these conditions, it becomes even more logical to turn away from the profession and turn instead to family, friends, and others, in order to seek the terms of a new identity. Cooling-out represents a socially induced, but individually managed means by which careers are controlled in science.

CAREERS IN OTHER ACADEMIC FIELDS

A study of academic physicists' careers naturally begs the question: to what extent do the observed patterns pertain to practitioners in other scientific fields and in other academic fields more generally? To answer the question, it is important to consider both the universalities and particularities of physics.

With regard to universalities, one may take note of the fact that these physicists are *academic* physicists whose careers have been studied within the contexts that structurally and culturally situate the system of U.S. higher education institutions. Their careers are substantially structured and acquire significant meaning by way of reward systems of universities. Reward systems of universities apply as much to physicists as to classicists, sociologists, biologists, and philosophers. The elite, pluralist, and communitarian academic worlds are not outcomes merely of physics or physicists, but of a system of meaning about careers as situated in an array of university types. It does, of course, take more than physicists to create such a system of meaning; such a system is created by a broad array of people in fields throughout universities to create a comparative understanding of what it means to be situated at a given point on the institutional continuum.

Because elite, pluralist, and communitarian identities are organizationally based, it is logical to conclude that these worlds of work are inhabited by practitioners in all academic fields (proposition 20). As organizationally bound constructs, these collective identities convey general sets of conditions of institutional life. There are physicists at elite universities just as there are philosophers at elite universities. There are physicists at pluralist and communitarian universities just as there are sociologists and chemists at pluralist and communitarian universities. The reference groups that these physicists selected to guide and make inferences about their careers are availed organizationally, in light of the various missions of the university types. There is nothing to suggest that these reference groups or their differential manifestation or usage across different types of universities is unique to physicists. Sociologists in pluralist institutions can die in research and become all-star teachers just as physicists can, or enact the variety of other career patterns observed in the pluralist world. Likewise, sociologist researchers can be blunted in communitarian institutions and continually pressed in elite institutions, much like physicists.

Since these identities are organizationally based, and because they convey general sets of institutional conditions for careers, one may deduce that, in general, careers in varieties of fields will transpire in roughly equivalent fashion as those found here. The major point: in all fields, academic careers are typically begun with high expectations. High expectations are compatible with fulfilling institutional goals of science, which are parallel to the institutional goals of any academic field in higher education—to extend socially certified knowledge. People normally do not embark on an academic career, in physics or in music, with the goal of being an also-ran. High expectations, and their evolution, situate careers not just in physics but in all academic fields. Marked by high expectations, particularly in the early phases of a career, across the three prototypes of American universities, careers may be seen to unfold in roughly similar ways (proposition 21). This is a testable proposition that may be subjected to future data on the rhythm of university-based careers in other fields.

In the introduction, physics and physicists were discussed for their exemplariness. Much was made about the pantheon of great physicists who inspire those who follow in physics to pursue exemplary careers. The point was made that the pantheon of physicists is perhaps the best known in the popular mind. Would not this reality differentiate careers in physics from those in other academic fields?

While the field of physics occupies a special location among fields as one that attempts to answer metaphysical questions empirically, this does not necessarily mean that the structure of careers in physics is systematically different from that of other fields. All academic fields, and perhaps all professional lines of work requiring scarce levels of talent, such as sports, medicine, or architecture, have a pantheon—a circle of mythic-like individuals whose achievements set a superior standard. Thus it is not that physics and physics alone has a pantheon to inspire high career expectations for achievement. All fields do. It is that these other circles are less widely apparent in popular culture than physics.

Whether or not pantheons in the fields of education or political science or botany are known widely, they are known (albeit in perhaps uneven ways) by practitioners in those fields, and, as a reference group, they help to establish normative career expectations. Hence, *just as physics careers are organizationally and professionally bound, so, too, are careers in other academic fields. Both internal/organizational and external/professional systems of reward govern performance across fields* (proposition 22). These universal arrangements further point out the conditions in which observed career patterns may hold generally in and out of physics. So much for universalities.

What of the particularities of physics that may make career patterns of physicists distinct from those of other academic fields? A source of possible answers lies in the phenomenon of *codification*, examined by sociologists of science and higher education researchers. Codification refers to "the consolidation of empirical knowledge into succinct and interdependent theoretical formulations" (Merton and Zuckerman 1973, 507). Academic fields and specialties within them vary in their extent of codification. Generally, highly mature fields are said to be highly codified; less mature fields, less codified. Comparatively, fields such as physics and chemistry are recognized as highly codified. Fields such as sociology and history are recognized as weakly codified. Fields such as psychology and biology may be recognized as possessing an intermediate degree of codification (for expanded discussion of codification, see Braxton and Hargens 1996; Hargens 1975).

An important aspect of codification is *consensus*, the extent to which practitioners of a field agree (Cole 1983; Cole, Simon, and Cole 1988; Hargens 1975; 1988; Hargens and Kelly-Wilson 1994; Zuckerman and Merton 1971). Agreement may be understood to have many referents: problem choice, methods for research, theory selection to explain phenomena, and the like. High-consensus fields are characterized by practitioners with a high level of agreement about what problems are worth researching, why they are important, the methods to be used to research them, the theories to be used to explain research findings, and so on. Low-consensus fields are marked by disagreement about which problems are most important to research and which methods and theories should be used in the research. In low-consensus fields, there may even be a greater mix of methods, some less well-developed than others. Theory may on the one hand be numerous and scattered, but on the other hand substantially uncultivated.

One can surmise that another referent of agreement consists of a collective definition of career success. That is, the extent to which members of a field agree on what constitutes a successful career and on which members of the field are successful, given the qualitative and quantitative characteristics of their achievements. Following this logic, one would expect physics, as a high-consensus field, to possess members with relatively clear and delimited definitions of success in the field. Correspondingly, one would expect low-consensus fields, such as sociology and history, to possess members with relatively ambiguous and varied definitions of success in their respective fields. That physicists (and scientists in other highly codified fields) say that they can be rank ordered by achievement is testimony of high consensus about success norms in the field.

> Physics is a field in which there is a rather fixed pecking order. People will agree that people's achievements can be rank ordered. Universities' and departments' achievements can be rank ordered and there won't be much argument. You know where you stand. It's sort of glaringly obvious. (54L)

That, for example, sociologists would say that it would be quite difficult to rank members of the field by achievement is testimony of low consensus about success norms in the field.

What does this mean for career patterns? It means that in high-consensus fields, the size of the elite is likely smaller, since fewer people can satisfy a stricter standard of success. By contrast, in low-consensus fields, the size of the elite is likely larger, since more people can satisfy either a weaker standard or one of numerous standards that are employed to gauge success—success meaning widely different things to different people.

If this is true, one would expect members of high-consensus fields, such as physics, to offer among the severest judgments about their careers. One would expect members of low-consensus fields to find the greatest latitude in the judgments they could render on their careers. Put differently, members of low-consensus fields have more chances to define themselves as successful because they can more easily find a sanctioned reference group against which they favorably measure up. Members of high-consensus fields have the chips stacked high; career success hinges on an ability and opportunity to satisfy relatively rigid collective understandings of achievement.

One might also therefore predict that low-consensus fields offer the greatest opportunity for professional satisfaction; practitioners can do almost anything and find an outlet to be recognized for it. It is not difficult to think of fields that exhibit this quality. Professional satisfaction in high-consensus fields is a scarcer commodity, since it is traded for scarcer talent. These formulations point out a further irony: *the chance of disappointment is greatest in fields with the clearest collective minds, whereas the chance of disappointment is lowest in fields in disarray* (proposition 23).

The *resource dependence* of physics is another means by which it achieves particularity. To do their physics, physicists need money. Money is necessary for numerous components that comprise research in physics: laboratories, equipment, supplies, staffs of post- and predoctoral researchers, professional travel, release time from teaching, and an array of indirect costs. A scientist helped to place the career realities of physicists in the following terms:

> [I]t would be easier to have a little more funding and not have to worry about that. That is the real mental burden, if you ask anybody what things contribute to their anxiety level. Not knowing about funding is probably the biggest thing, because that's just not in your control. I can't control the students that walk into my lectures, but I can work with them, and I can learn their personalities, and I can do what I can do. That's under my control. But the funding is really hard, and you need it to get stuff done. If you need liquid helium to cool your experiment and you don't have money, you can't get liquid helium, you can't cool it down, and you can't do the measurement. This is a real problem. It's not just, "Well, it's an inconvenience, so use liquid nitrogen, it's cheaper." If you've got to get to 4 degrees, you've got to have liquid helium, and it costs real money. So one is worried about "Can I get this experiment done? Can I get this student through? Will my grant get

renewed? Will I have the money to finish this project?" It's a real worry. ... You wake up in the middle of the night and worry about stuff like this. (48L)

Not all academic fields, of course, are resource dependent in the same ways or degrees. Other fields in the hard sciences, such as chemistry and biology, will approximate conditions of physics. Fields in the humanities are significantly less resource dependent or relatively resource independent. Fields in the social sciences compose a mix of resource dependence and independence. For example, anthropological work that relies on data obtained in distant field sites carries greater resource demands than sociological work that relies on observational data obtained on inner-city street corners. There are also variations by specialty area within fields, some more resource dependent than others. Experimental social psychology, for instance, imposes greater resource demands than most research in the sociological study of social movements. Moreover, theorists in all fields are less resource dependent than experimentalists or other types of primary empirical researchers.

In addition, academic fields differ in their *mutability*, that is, the capacity of a researcher to change direction or research area entirely to a less resource-dependent project should a more resource-dependent line of research fail. Relatively speaking, sociology, for example, is highly mutable. In the absence of funding, most sociologists can turn to other projects that are less resource contingent, and often may be able to do so with few or no career costs. By contrast, physics is relatively immutable. Virtually all physics research, save a fraction of purely theoretical work, is resource dependent. Doing physics of almost any kind requires a significant financial infrastructure.

Together, *the high resource dependency and immutability of physics establish notable constraints on academic careers in that field and in fields like it* (proposition 24). One would again expect practitioners in such fields to offer the severest judgments about their careers because, when these contingencies fail, the consequences for careers are likewise severe. But even when contingencies remain intact, the risk and anxiety about their collapse remain high, since practitioners can easily anticipate the consequences of failure. Even in good times, one is apt to find physicists (and academics like them) on edge because everyone knows money will run out at some point, and sometimes prematurely, and must be renewed through successive rounds of highly competitive and taxing grant application. As one recalls, this process alone made many physicists question whether they would again seek an academic career.

Furthermore, *if a field is highly resource dependent, then the function of cumulative advantage and disadvantage assumes particular significance* (proposition 25). It means that when resource contingencies fail, it can spell dire consequences for careers as disadvantages begin to accumulate. It is very difficult in a field such as physics to obtain research funding after a long dry spell, and by *long*, one is talking on the order of just about five years. On balance, highly successful physicists are more or less constantly funded. This is an aspect of the field's immutability. Success hinges on external support. In its absence, the process of cumulative disadvantage severely punishes many competent practitioners by disqualifying them from future rounds of significant research endeavor.

By contrast, low resource dependent and mutable fields establish notable flexibility for academic careers. Success does not hinge to the same degree on external support. Even if it does in certain instances, a field's mutability allows practitioners to change course to less resource-contingent ventures, and the career proceeds. A career that proceeds productively is more likely subject to cumulative advantages, not disadvantages. Low resource-dependent fields may also employ different meanings for low productivity. In fields like sociology, in contrast to physics, it is more possible to obtain research funding after a dry spell. The cognitive order of such fields may require time to develop fundable projects. Humanists, for example, may sometimes "wait for ideas to come." A project may take many years to develop prior to actual research. A gap in publication does not necessarily mean failure in English as much as it may in physics. Hence, practitioners in such fields are again exposed to greater opportunity for success. This is the case not only because a field may be weakly codified, opening up what success means and making it easier to meet definitions of it in a given set of ways, but also because of a relative lack of contingencies.

One would also therefore expect the careers of women in low- and mid-consensus fields to assume a wider variety of patterns, in contrast to that observed among the subsample of women physicists, who uniformly displayed elite patterns. On the one hand, success in low- and mid-consensus fields can be defined in varieties of ways, and thus there are varieties of career performances that can satisfy a definition. On the other hand, low- and mid-consensus fields will tend to be populated by more women to begin with, thereby potentially offsetting the highly conservative success thresholds for women in fields that are sparsely populated by them. For these reasons, *one might expect practitioners in resource-independent and in mutable fields to offer less harsh judgments and more sanguine perspectives about their careers* (proposition 26).

A final issue to consider consists of selection effects. That is, a counterargument holds that the observed differences in people's careers among the academic worlds is attributable not to the conditions of those worlds and how they socially shape careers, but to the choices, or selections, that individuals make about where to work. I took up the issue of selection effects in detail in the foundational work (Hermanowicz 1998, 131–137). There are several reasons to believe that effects of selection are minimized in the observed results.

First, choice is constrained by a labor market. In the present instance, the labor market of academia is, and has been since the early 1970s, highly competitive. Only a small fraction of people who complete doctoral programs end up obtaining academic employment. Although the ratio of job applicants to the one individual who is successful in a search varies by field, the ratio tends to be very high in present times and dating back many years. For example, in English, the ratio can be 150 to one or higher. In French, mathematics, sociology, and physics, among many other fields, the ratio is often as high or higher. It would be safe to say that more academic fields are marked by high ratios than fields marked by low ones. Under such conditions, most people, most of the time, go where they get a job offer, rather than select among numerous options. Those who would have been in a position to exercise greater agency in where they took a job are those academics who received their doctorates prior to 1970, before academic labor markets tightened in the aftermath of the great expansion of U.S. higher education in the 1950s and 1960s, a group of people who correspond to the eldest cohort in the present work.

Second, in the foundational work, direct comparisons were made between scientists who received their doctorates at roughly the same time and from roughly similar institutions in order to see how their perceptions of work and career varied by virtue of the different institutions in which they obtained academic employment. The evidence indicated quite clearly that, despite having similar doctoral origins and similar aspirations as graduate students for their anticipated careers, their perceptions of work and career diverged dramatically in light of the vastly different types of institutions where they obtained a professorial position. Graduates of the University of California–Berkeley who obtained a position at, for example, the University of Tulsa, perceived the conditions of their careers in terms starkly different from similar graduates who obtained a position at, for example, Princeton. They did so not because they really wanted to work at Tulsa and not at Princeton, but because the conditions at Tulsa were unaligned with expectations and aspirations for a career that they had been socialized to develop in their doctoral training.

Third, the present work, in studying the same people at two points in time, provides the opportunity to assess the ways in which a work context does, or does not, affect understandings of one's career. Longitudinal study of this sort sets a stage to see how work contexts shape thought and behavior. The evidence suggests strongly that people's perceptions of their careers vary so significantly not simply because of choices they made decades ago about the institution where they would work, believing those given institutions would be most compatible with their expectations for an academic career, but because people have had to make substantial adaptations to their expectations in light of the conditions posed by their employing institutions. The preponderance of evidence supports the existence and bearing of social effects of environments on the shaping of careers over time as opposed to effects of individual self-selection at one particular point in time.

FUTURE COHORTS OF SCIENTISTS AND CONTEXTS OF SCIENCE

To account for the career patterns observed, I have turned to the expectations that scientists have for their careers. My consideration of anomie and cooling-out identified social-psychological symptoms and means of individual adaptation when expectations are incongruent with the opportunity and ability to meet them. This has implied that expectations *exceed* opportunity and ability. But what about when career expectations are *low*? Do some scientists begin their careers with low expectations? And are they thus more prone to greater congruency with their ability and environmental opportunities to succeed?

One can be certain that expectations for careers, in science as in numerous other lines of work, are not of one kind, once the career gets underway. The evidence clearly indicates that not all scientists across all worlds of science pursue the same goals or do so with the same intensity and commitment. Rather, the nature of individual goals and individual intensity and commitment is an expression of social structure; sociologically, individual expectations are but the manifestations of contextual opportunity and constraint, thus helping to give rise to the distinctive identities of elite, pluralist, and communitarian scientists. But what of expectations at the time of career onset, just prior to putting these career processes into play?

The evidence on hand from both the foundational and longitudinal studies suggests that for most scientists starting-out expectations are customarily very substantial. I point to three main reasons why this is so. First, scientists at the time of career onset are a highly selected group, having been filtered through a long process of trial and competition to achieve positions in the professoriate. This lengthy process generally selects for, among other things, an intense motivation and commitment to work and a strong desire for success in science (Simon 1974). Put differently, a process of cumulative advantage and disadvantage is already underway, having positioned these scientists (who in fact entered scientific careers) more beneficially than those who were refused or who otherwise could not gain academic employment.

Second, science is a profession. As such, it is culturally construed and maintained as an arena in which individuals are to make significant achievements (Bledstein 1976; Haber 1991). People who enter professions tend to be achievement-oriented (Hernstein and Murray 1994; McClelland 1961; Paul 1980). To enter a profession lame and to wish to do little in it runs counter to both socialization and training for it and its social mandate for members.

Third, scientists have a pantheon. Whether budding at age fifteen or wilting at age fifty, scientists recognize the pantheon. They feel its influence, know its power, and are simultaneously inspired and humbled by its charisma, even as its salience varies from one world of science to another and over time in a scientific career. Beginning a scientific career with low expectations is thus inconsistent with the institution of science—to extend socially certified knowledge through research and discovery.

What if scientists begin their careers in institutions that are not research-oriented, wherein their individual expectations may thus come to be mapped by the (counter-research) expectations of the institutional social structure? All of the institutions in the present sample expect research from their faculties, albeit in varying degrees. This, however, was not always the case. The communitarian institutions in the sample have embraced research in greater degrees compared with when the eldest cohorts in the sample entered them. *If one were to find scientists (and other academics) who began their careers with more modest career expectations, it would tend to be in communitarian and in some pluralist institutions, which have since altered their missions and become more research-oriented, and then in the eldest cohorts of faculties in those institutions, who began with a less research-oriented emphasis* (proposition 27). Chronologically, these would tend to be faculty members who obtained both academic employment and tenure prior to around 1970, a generation that at this writing is in the middle of the process of retiring.

In such cases, where careers begin more modestly, expectations are comparatively easier to satisfy, at least at the beginning of the career. As their careers unfold along what is typically a greater teaching orientation, and as organizational missions change in favor of research, these individuals become anachronisms. One finds such cases of individuals, where they occur, in late career decoupled from their departmental and institutional environments, since their career orientations are long since out of synchrony with institutional expectations for careers. A handful of these cases were described in the foundational study, when the individuals were then in late career. These scientists' outlooks tended to be mixed. They identified with their original career orientation, with its emphasis on teaching, students, and local institutional service, but had developed an ambivalence about their institution and colleagues who were more research-minded. They typically sensed that they have been "shoved aside." They were the ones whom more research-oriented colleagues, in their own interviews, occasionally discussed and asked aloud: "When are they going to retire?" (See Hermanowicz 1998, 151–152.) Now in post-career phases, these scientists experience the withdrawal (typical of pluralists) and the separation (typical of communitarians) from the career, any antagonism toward the scientific reward system now essentially gone.

If, with the presently aged and arrayed sample of scientists, modest expectations are substantially less common at career onset than high expectations, what may the future hold for cohorts in and contexts of science? A glimpse is provided by what the scientists themselves had to say about academic careers and the prospect of entering one again, results of which are presented, by career phases, in table 33.

Scientists are decidedly mixed about whether they would again seek an academic career. Only elites in early- to mid- and in mid- to late-career phases definitely would. Elites in late- to post-career phases would, but they are less emphatic than their younger counterparts. The greatest plurality of response is observed among pluralists; some say "yes," others "no," and still others "maybe." Communitarians are most ambivalent about the prospect. The ambivalence begins in early career and for many lasts as long as the scientists are living. In short, *there is far from definitive and emphatic enthusiasm for academia among those who have forged careers in its various institutions.* The

TABLE 33. Scientists' perspectives on seeking an academic career again, what scientists would do differently, prominent concerns, thoughts of leaving present institution, and thoughts of retirement, by career phases

SCIENTISTS' PERSPECTIVES ON SEEKING AN ACADEMIC CAREER AGAIN

Phases	Elites	Pluralists	Communitarians
Early to Mid	Definitely	No; Maybe; Yes	No
Mid to Late	Definitely	Maybe; Yes	No; Maybe
Late to Post	Yes	Maybe; Yes	No

WHAT SCIENTISTS WOULD DO DIFFERENTLY IN AN ACADEMIC CAREER

Phases	Elites	Pluralists	Communitarians
Early to Mid	Very Little; Nothing	More Research Focus; More Aggressive	"Better" Institution
Mid to Late	Nothing	More Mentoring; Nothing	Different Institution
Late to Post	Technical Decisions	Different Field	Different Field/ Different Institution

SCIENTISTS' PROMINENT CONCERNS

Phases	Elites	Pluralists	Communitarians
Early to Mid	Funding	Future Role; Commitment;	Professional Opportunity Funding
Mid to Late	Funding; Time	Career Direction or Funding	Retirement; Boredom
Late to Post	Professional Standing	Retirement Timing	None

(continues)

most positive responses come from elites, yet comparatively few academics will ever be employed by elite institutions (proposition 28).[6]

The relative positive response among elites is reinforced by what they said they would do differently were they starting their academic careers over again. Most elites would do very little differently. The eldest among them would reconsider technical decisions that they now see weighing on the extent of their achievement and recognition. In early- to mid-career phases, as they find themselves seeking career direction, pluralists state they would seek to develop a greater research focus and more aggressively develop a

TABLE 33. (continued)

SCIENTISTS' THOUGHTS OF LEAVING PRESENT INSTITUTION

Phases	Elites	Pluralists	Communitarians
Early to Mid	No	Yes; No	Yes
Mid to Late	No	No	No
Late to Post	No	No	No

SCIENTISTS' THOUGHTS OF RETIREMENT

Phases	Elites	Pluralists	Communitarians
Early to Mid	No	Yes; No	Yes; No
Mid to Late	No	Possibly; No	Actively Contemplating; Already Retired
Late to Post	Retired; Planning	Retired; Planning	Retired; Planning

research orientation throughout the course of a career. Others among them at later career points would have sought more mentoring, again to better position themselves with respect to research and its potential rewards. Still others would seek a different field. Mirroring their ambivalence about seeking an academic career again, communitarians, if they did so, would make more radical changes. They most typically state that they would seek "better" or different institutions, by which they mean those more conducive to scientific research.

In their careers, elites are most concerned about funding and time with which to fulfill their numerous responsibilities. They are also always concerned about professional standing, though this achieves special prominence in latter-most career phases. Pluralists develop concerns about their commitment to science, the direction of their work, and funding, but ultimately supplant those with the question of when to retire. Much of the career in the communitarian world is taken up with concerns about the lack of professional opportunity. When much of the opportunity proves elusive, concerns turn to boredom and thoughts about when to retire.

Despite these multifaceted negative or ambivalent career sentiments, most scientists would not leave their present institutions. But this is a non sequitur. Scientists say this not because they do not want to leave, but because they cannot leave; they have incurred too many sunk costs, which disqualifies them from seeking an alternative career.

They would, however, more readily entertain retirement, a finding prominent among pluralists and communitarians. In pluralist and communitarian worlds of science, it is easy to find scientists at any point in their careers who say they would retire if they could maintain their present standard of living. This attitude runs counter to an excitement and corresponding commitment that one might otherwise find among scientists. The evidence indicates that in those worlds of science that employ the greatest number of scientists—the combined worlds of pluralists and communitarians—significant fractions would readily leave if able.

The picture that emerges is far from sanguine. It is difficult to take this evidence, and that presented throughout this and the preceding chapters, and argue unabashedly about the satisfaction scientists find in pursuing, and in having pursued, an academic career. In what direction do the patterns seem to be headed? Research on academic institutions offers a first step toward an answer.

Alternately called "mission creep," "academic drift," and "institutional upgrading," the increasingly widespread phenomenon in which institutions of many types seek to embrace the model of the American research university has become a subject of higher education research (Finnegan and Gamson 1996; Henderson and Kane 1991; Morphew 2002; Neave 1979). The research emphasizes institutional benefits derived from this status change, including enhanced status and prestige that in turn can marshal additional resources, such as attractively credentialed faculty, students, and monies from legislatures, foundations, and other funding agencies; greater program offerings and correspondingly greater market shares of students; and increased tuition revenues and alumni giving. Often a change in name occurs as part of this evolutionary process; colleges become universities, highlighting the claim to new status and the aspiration to command new resources (Brint, Riddle, and Hanneman 2006; Dunham 1969; Ruch 2001). This generic procedure is as common to organizations as to occupations. Cashiers become sales associates; mechanics become automobile technicians; strippers become exotic dancers. Among academic organizations as among occupations, the change lays a claim to more rarified status, bespeaking a more esoteric and skilled activity, and hence entitlement to command scarce rewards. One of the communitarian universities in the sample changed its name once and a second has done so more than three times since its founding. The third communitarian institution in the sample, a teachers' college at its founding, has similarly changed its name and now declares itself "one the nation's top-100 public research universities" in its publicity brochures.

By one view, these substantial changes in the organizational makeup of higher education institutions in the U.S. may spell greater research opportunity for individual scientists and other faculty members than existed within the population of institutions at a prior point in time. This remains an empirical matter that merits systematic treatment. *While there may be a positive net change in research opportunity, this of course does not mean that there is congruency between the research expectations of individual scientists and those of their employing institutions. Where the former exceeds the latter, one may expect to witness the same cooling-out processes as found endemic throughout the present work (proposition 29).* This prompts a more general point.

While one cannot safely conclude that mission creep brings about greater research opportunity, one can say safely that it does entail a change in institutional expectations for careers. And the change, unsurprisingly, involves a greater emphasis on research productivity. This evolutionary process toward a more intensified stress on research has taken place amidst other changes in the institution of science in particular and higher education in general. Scientists of all generations note a heightened competition for research funding. Pressures to publish are now more intense as tenure and promotion procedures have grown more formalized throughout the higher education system, and as the supply of labor replacements has increased, making it easy to substitute faculty members whose records prior to tenure may be deemed good, but not good enough to satisfy present-day performance realities, as the comparison among tables 20, 23, and 26 made apparent. Awards have proliferated, noting not simply worthy scientific work but also the scientific community's increased emphasis on awards (Zuckerman 1992). This pattern appears typical throughout most, if not all, academic fields. For example, in sociology, where once there were fewer than a half dozen awards given annually for performance deemed exceptional, there are now awards given by every section of the field, for articles, for books, even for student submissions. These conditions, already having become or well on their way to becoming institutionalized as to enter habits of thought and behavior, have altered what it means to lead an academic life.

Taking into account the longitudinal evidence of the study, organizational changes in institutions documented in the research cited above, and the omnipresent scarcity of rewards, the following proposition is drawn:

increased emphases on research will be accompanied by increased probabilities of anomie throughout the system of higher education (proposition 30). As research is more greatly stressed, by institutions as by individuals, career expectations rise, in accord with attempting to satisfy external reference groups that are consistent with fulfilling the goals of science. As expectations rise, the likelihood of satisfying them decreases, because the expectations are defined by that not yet achieved and, ultimately, by the unachievable. These conditions favor dissatisfaction and disaffection for the academic career, much as was found among the many scientists who would seriously question seeking one again.

Thus, as institutions increasingly embrace the research model of the university, careers will increasingly experience pressure to conform to the model. This is by no means to say that careers shall become monolithic. The reward system will always operate, and, as the research literature indicates, individuals will reap rewards at a rate that progressively deprives others who are unrewarded. Institutions, like those present in this study, will retain their special emphases because they have a public mandate to do so and because, by definition, only a finite set of them can constitute an elite, a certain small set of career performances set apart socially from others. But research will, as it does increasingly now, constitute a more prominent component of roles in institutions outside of the elite, which strive to become more like it. With expectation for success in research more widespread, conditions form for the spread and intensification of anomie in academe.

It will be incumbent upon future work to further trace the trends of how scientists, and other academics, perceive their careers. Such perceptions tell us not only about individuals but about institutions. How will universities be organized, with what expectations and systems of reward? And how, then, will academic careers evolve—in another ten, fifteen, or twenty-five years? An opportunity is presented to see how self and career continue to change as the institutional and organizational terms of academic work evolve over time. At stake is the profession. It will be important to see which types of people, with what levels of talent, the profession is able to attract to an academic career. One scenario is that the profession will attract less talented individuals. More talented individuals, seeing the conditions under which academic careers are experienced, tight job markets, heightened difficulty of tenure and promotion, and the general scarcity of reward compared to investment of time and effort, appear increasingly to be entering other professions. In the case of physics, Stephen Cole has offered preliminary evidence that this is already the case:

> [T]he most important reason for the observed decline in the number of talented young physicists is the significance of opportunity structure in influencing the career choice of talented individuals. People who select science as a career are obviously those who have an intense interest in the subject. At least in our times, however, intense interest does not seem to be enough. Many potential creative scientists are discouraged by the lack of secure employment prospects. As the job market for academic scientists has tightened, the attractiveness of the occupation to talented as well as less talented youth has decreased. Whereas in the 1960s talented young people may have selected the sciences in preference to other professions or business, that trend seems to have been reversed. (Cole 1992, 226)

News outlets in the popular media have caught on to at least some parts of a social problem. Headlines have begun to broadcast a version of the news: "The real science crisis: Bleak prospects for young researchers" (Monastersky 2007). "Young scientists hit the hardest as U.S. funding falls," and so on (Hiles 2006).

It is conceivable that less talented individuals would possess lower expectations for superior achievement, thus muting the effects of anomie and leaving them more contented with work. But such a net effect would have to overcome the effects of induction, training, and socialization, the power of the pantheon in inspiring peak performance, and, of course, scientific norms that press for productivity. These form the conditions of a profession and of the changing personal perspectives, shaped in specific settings over time, on what it means to have developed a career in science.

APPENDIX A
Interview Protocol—Foundational Study, 1998

INTERVIEW OF SCIENTISTS

This is a study about the aspirations of academic scientists. The questions I would like to talk about deal with one's individual identity and how that identity has unfolded over time. Some of the things I will discuss ask you to reflect upon yourself and often involve making personal judgments that will touch on various professional and related personal topics. Your participation in this study is strictly confidential. Interviews are normally tape-recorded, and this simply provides for accurately keeping track of information. Subsequently the tape will be destroyed. Your participation in this study is important. However, should you at any time wish to stop, you may do so without prejudice to you, and at any time you should feel free to ask me questions concerning the interview or the study. May we begin?

A. LOCATION IN THE DIVISION OF SCIENTIFIC LABOR
  1. Can you describe the type of work you do?
  2. To what extent is your work collaborative?
  3. [If collaborative] How large are the collaborative teams on which you work?

B. CONSTRUCTION OF PERSONAL HISTORIES AND PERSONAL IDENTITIES
  1. What aspirations did you have as a graduate student?
    Probe: What did you want to attain?
  2. In everyone's career there are "roads not taken"—different avenues you might have followed. What have been the ones for you?
  3. What consequences have these outcomes had on your career?
  4. How did you come to arrive at this university?

5. You were a graduate student at _____ Is this the type of university where you wanted to end up?
6. How have your aspirations unfolded since being a graduate student?
7. How has being at this university affected your career?
   Probe: How has this university constrained your career? How has it helped your career?

C. GENERALIZED DEFINITIONS OF SUCCESS LADDERS
   1. What do you associate with a "successful" career in physics?
   2. What do you think are the most important qualities needed to be successful at the type of work you do?
   3. What does *ultimate* success mean to people working here?
   4. Is there an understanding of a *minimum* needed in order to maintain respect among people here?
   5. Is there an understanding of a *failed* career among colleagues here?
   6. Taking your colleagues in this department, how would you say their success varies?
      Probe: Have they advanced at the same rate?
   7. Where do you place yourself among that variety?

D. CONCEPTIONS OF FUTURE AND IMMORTALIZED SELVES
   1. What do you dream about in terms of your career?
   2. What ultimate thing would you like to achieve?
   3. How do you envision yourself at the end of your career?
   4. How would you like to be remembered by your colleagues?
   5. What about your life do you think will outlive you?

E. AMBITION
   1. Would you say that you are ambitious?
      Probe: Would you say that you have a strong will to succeed?
   2a. [If yes to 1] What is your ambition?
   2b. [If no to 1] Would you say that you have a strong will to succeed?
   3. Where does your ambition come from?
   4. What role do you think ambition plays in your life?

F. SELF-DOUBT/SELF-FRAGMENTATION
   1. What would you like to be better at?
   2. Has there been a significant time when things really did not go the way you wanted them to?

3. What major doubts have you had about yourself?
4. Have there been times when you felt that you let yourself down?
   Probe: Have you ever felt disappointed in yourself?
5. Has there been some inner conflict or turmoil that you have sought to understand in your life?

We are near the end of the questions I have.

6. I would finally like to ask about something you are most proud of. What stands out as something that has left a strong positive impression on you?

Source: Hermanowicz, Joseph C. 1998. *The Stars Are Not Enough: Scientists—Their Passions and Professions.* Chicago: University of Chicago Press, 211–213.

APPENDIX B
Contact Letter to Scientists

April 7, 2004

Dr. ____
Department of Physics

____

____

____

Dear Professor ____:

It has been 10 years since we met. In 1994, I interviewed you for a study of careers in science. Funded by the National Science Foundation, and conducted under the auspices of the University of Chicago, that study explored scientists' aspirations and identities related to their work. The study was based on interviews with scientists across the United States. As the principal investigator of that study, I know very well that you formed a critical part of the sample, and I remember very well how much your participation contributed to the work.

I write to ask for your help. A 10-year follow-up study is being conducted entitled *Lives of Learning: Continuity and Change in Science Careers*. The study design calls for interviewing the same participants who composed the original work. This is both substantively and historically significant: the study will be the first of its kind to follow professors in their careers. It therefore holds real potential to generate important findings about how careers are experienced and understood by scientists themselves. The study presents the unique opportunity for you as a scientist to convey knowledge about careers and science acquired over the years of your extensive experience in physics.

Your participation would involve an interview, conducted again by myself, that would last approximately an hour. As before, interviews would customarily take place in your office, and I would meet you at an agreed upon time. (If you happen to be one of the several scientists in the sample who has retired, we would make alternative meeting arrangements as necessary and as agreeable to you.) The interview would consist generally of questions about changes and continuities in your career over the past 10 years. Like before, the interview would normally be tape-recorded simply to keep accurate track of information, and subsequently the tape would be destroyed once the study is completed. Participation and all interview material will be strictly confidential. Both personal and institutional identities will be concealed in published work, following standard conventions of work of this kind. Participation is voluntary. Nevertheless, I very much hope you can participate; the success of the work depends on you. All aspects of this project have passed the usual human subjects reviews at the University of Georgia.

I will call you shortly to invite your participation and answer any questions you might have. Please know how greatly I appreciate your time and help with this request.

Yours sincerely,

Joseph C. Hermanowicz
Assistant Professor

APPENDIX C

Thank-You Letter to Scientists

May 15, 2004

Dr. _____
Department of Physics

_____

_____

_____

Dear Professor _____:

Having recently completed our interview, I want to take the opportunity to thank you for all your help. You are most kind and gracious not only in your time, but most especially with your insight about careers in science, and your capacity to communicate some of the meaningful aspects of your life in physics. This means more to me, and to the work it forms, than I can tell.

If you have any questions or want to get in touch with me, you should feel free to do so at any time. My departmental address and telephone number are on this letterhead, and my e-mail address is: _____.

As you know, this project has passed customary human subjects reviews. Should you have any questions regarding your rights as a research participant, you may contact me or_____, Human Subjects Office, University of Georgia, 606A Boyd Graduate Studies Research Center, Athens, GA 30602-7411; Tel: (706) _____; E-mail: _____.

Please accept my many thanks for all the help you have given, and my very best wishes for the months and years to come.

Yours sincerely,

Joseph C. Hermanowicz
Assistant Professor

APPENDIX D
Interview Protocol—Longitudinal Study, 2004

Date: _____ Respondent: _____

_____

_____

_____

Notes: _____

_____

_____

_____

_____

_____

_____

LIVES OF LEARNING: CONTINUITY AND CHANGE IN SCIENCE CAREERS

Interview Protocol

This is a study about continuity and change in academic careers. The questions I would like to talk about deal with one's personal identity and how that identity has unfolded over time. Some of the things I will discuss ask you to reflect upon yourself and often involve making personal judgments that will touch on various professional and related personal topics—much as we discussed 10 years ago. Your participation in this study is confidential. As before, interviews are normally tape-recorded, and this simply provides for accurately keeping track of information. The tape will be destroyed at the completion of the study. Your participation in the project is vital to its success in learning about academic careers and tracking how people progress and age in them. However, should you at any time wish to stop, you may do so without prejudice to you, and at any time you should feel free to ask me questions concerning the interview or the study. May we begin?

Several of the questions ask you to respond in the context of the 10-year time frame that has elapsed since we last talked. There may be markers in your life and work that you may think of to help bring this span of time fresh into mind.

For example, 10 years ago, according to information you gave me then, you were about _____ years old. You had been at _____ for about _____ years. As we talk, try to conjure how the intervening years have unfolded.

Time Start: _____
Time End: _____

CV: *Yes? / No? / Arrangements?*

I. CHANGES AND CONTINUITIES
   1. What changes have you seen in your perspective on work?
      Probe: What changes have you seen with regard to research?
      Probe: What changes have you seen with regard to teaching?
      Probe: How has your outlook on your career changed?
      Probe: How would you complete the sentences: "I am more X"; "I am less Y"?
      Probe: How have you *not* been successful?
      Probe: [For those at institutions different from 1994] How did you arrive at this institution?
   2. What have been the most significant changes *in your life* outside of your career in the last 10 years?
      Probe: How do believe this has affected your career?
   3. Looking back over the past 10 years, has your career progressed as you expected?
      Probe: If yes: How so?
      If no: Why not?
   4. Do you think you are working harder, less hard, or about as hard as you were 10 years ago?

5. Would you say you are more ambitious, less ambitious, or about as ambitious as you were 10 years ago?
   Probe: Why more, why less?
6. How do you think academic careers are different for those older and younger than you?
   Probe: Easier? More difficult? More competitive? Less exciting?
7. How do you think academic careers are different at higher and lower ranked departments?
   Probe: Do you think people age in their work differently?
8. Do you believe you define success differently now than you did 10 years ago?
9. If you were a university president, what one change would you make?

II. SATISFACTIONS

10. On a scale of 1–10 (10 being strongest), how satisfied with your career would you say you are?
    Probe: Why is it x and not 10? What would make it a 10?
11. At what age would you say you have been the most satisfied in your career?
    Probe: Why then?
12. Over the past 10 years, what would you say have developed as your 3 biggest joys about your job?
13. In learning what you have about academic careers, would you go into an academic career if you were starting all over again?
    Probe: Why?/ Why not?

III. DISSATISFACTIONS

14. If you were starting all over again, what would you do differently, knowing what you know now about your line of work?
    Probe: Why would you do those things differently?
15. Over the past 10 years, what would you say have developed as your 3 biggest complaints about your job?
    Probe: If you go home and complain about something, what would it be?
16. What worries or concerns do you have about your career?
    Probe: Do you feel sufficiently recognized for your work?
    Probe: What gets you down?
17. What frustrations have you had?

18. Have there been ways in which an academic career has been unrewarding?
19. Over the past decade, have you seriously wanted to leave this university?
    Probe: Why?/Why not?
20. In thinking about your graduate training, what do you view as its greatest weakness or deficiency?

IV. ASPIRATIONS

21. In light of the past 10 years, what are your current aspirations?
    Probe: You have a good —— or more years to go. What do you want to do?
22. What dreams do you have in relation to work?
    Probe: What would you like to attain?
23. How have these aspirations changed over the past 10 years?
24. [For those not retired] If you could retire now, and lead approximately the same life you have currently, would you?
    Probe: Why?/Why not?
    [For those retired] Since retiring, what do you miss the most about your job?

V. RETIREMENT SUPPLEMENT
   (QUESTIONS ONLY FOR THOSE RETIRED)
   A. For you, what has been the worst part about retirement?
      Probe: What types of adjustments has this required?
   B. What, for you, has been the best part about retirement?
      Probe: What about retirement makes it desirable for you?

We are near the end of the questions I have.

25. I would finally like to ask about something you are most proud of. What stands out as something that has left a strong positive impression on you?

APPENDIX E
Post-Interview Questionnaire

LIVES OF LEARNING: CONTINUITY AND CHANGE IN SCIENCE CAREERS

Post-Interview Questionnaire

Information provided on this form is strictly confidential and will be used for research purposes only. At all times your identity will remain anonymous.

1. What is your current marital status?
   ( ) Married
   ( ) Separated
   ( ) Divorced
   ( ) Single

2. In the past ten years, have you:

	YES	NO
Divorced?	( )	( )
Remarried?	( )	( )
Had child(ren)?	( )	( )

   If YES, please list the year(s) of birth of all of your child(ren).
   1. 19___
   2. 19___
   3. 19___
   4. 19___
   5. 19___

3. Please indicate your past year's *nine-month base university* salary (or, if retired, estimate your annual income). (Do not include non-

university income, e.g., research grants, consulting fees, royalties, honoraria, etc.)
1. ( ) less than $20,000
2. ( ) $20,000–$29,999
3. ( ) $30,000–$39,999
4. ( ) $40,000–$49,999
5. ( ) $50,000–$59,999
6. ( ) $60,000–$69,999
7. ( ) $70,000–$79,999
8. ( ) $80,000–$89,999
9. ( ) $90,000–$99,999
10. ( ) $100,000–$109,000
11. ( ) $110,000–$119,000
12. ( ) greater than $120,000

APPENDIX F
Departmental Questionnaire

LIVES OF LEARNING: CONTINUITY AND CHANGE IN SCIENCE CAREERS

Departmental Questionnaire

Information provided on this form is strictly confidential and will be used for research purposes only. At all times, including in any published work, your identity and the identity of your institution will remain confidential.

Please answer the following questions as completely and as accurately as possible. Kindly return the completed form in the enclosed envelope. Your assistance in this research effort is greatly appreciated.

I. TEACHING
   1. What is the normal yearly teaching load for the average faculty member of your department?
      ( ) 1–2 courses ( ) 4–5 courses
      ( ) 2–3 courses ( ) 5–6 courses
      ( ) 3–4 courses ( ) More than 6 courses
   2. For the average faculty member in your department, what percentage of teaching in a given year involves general undergraduate courses?
      ( ) 75–100%
      ( ) 50–74%
      ( ) 25–49%
      ( ) Less than 25%
   3. What percentage of faculty members in your department teach mostly courses at the graduate level?
      ( ) 75–100%
      ( ) 50–74%
      ( ) 25–49%
      ( ) Less than 25%

4. Does your department have provisions for leaves from teaching?
   ( ) Yes   ( ) No
5. *If yes to Question 4*, explain briefly the conditions under which a faculty member can take a teaching leave.

6. *If yes to Question 4*, how often can leaves from teaching be taken?

7. As best as possible, indicate the extent to which teaching performance factors into promotion decisions in your department.
   ( ) A lot
   ( ) Somewhat
   ( ) A little
   ( ) Practically not at all
   ( ) Not at all

II. RESOURCES
   1. What is the amount of *annual federal support* to your department?

   2. What is the amount of *annual non-federal support* to your department?

   3. What is the *overall annual operating budget* of your department?

   4. Taking all physics departments in the United States, how would you characterize the present *equipment and research facilities* in your department?
      ( ) Among the very best
      ( ) Very good
      ( ) Average
      ( ) Fair
      ( ) Relatively poor
      ( ) Nonexistent
   5. What would you take to be the greatest problem(s) with your present equipment and research facilities? (Check all that apply.)
      ( ) Generally, they're old.

( ) Generally, they're limited.
( ) Generally, they do not allow for major research.
( ) They serve a relatively small fraction of the department faculty.
( ) Other (indicate below):
_____
( ) None of the above applies.

6. Do your new assistant professors receive university or department start-up funds or a similar allowance?
   ( ) Yes   ( ) No

7. *If yes to Question 6, what is the amount?*

8. *If yes to Question 6, do these funds have restricted professional uses?*
   ( ) Yes   ( ) No

9. *If yes to Question 8, explain what the funds may be used for.*

10. Do faculty members in your department have access to *departmental or university* funds for research?
    ( ) Yes   ( ) No

11. *If yes to Question 10, indicate the type(s) and amount of fund(s).*
    Type: _____   Amount: _____
    _____   _____
    _____   _____

12. Is each faculty member in your department covered for travel/conference expenditures with departmental or university funds?
    ( ) Yes   ( ) No

13. *If yes to Question 12, for how many trips/conferences per year? (Alternatively, explain the provisions.)*

14. Roughly how many employed *technical staff* are on hand for faculty members in your department? (Exclude postdocs and graduate assistants.)
    ( ) 1–5   ( ) 16–20
    ( ) 6–10   ( ) 21–25
    ( ) 11–15   ( ) More than 25

15. In a given year, roughly how many *postdoctoral researchers* have appointments in your department?
    ( ) 1–5   ( ) 16–20
    ( ) 6–10   ( ) 21–25
    ( ) 11–15   ( ) More than 25

16. Roughly what percentage of faculty members in your department have *graduate research assistants*?
    ( ) 0%
    ( ) 1–25%
    ( ) 26–50%
    ( ) 51–75%
    ( ) 76–100%
17. Taking all the faculty members who do *not* have graduate research assistants, what would you say is the single most important reason? (Check only one.)
    ( ) They do not need RAs because work is theoretical.
    ( ) They do not need RAs because not active in research.
    ( ) Pool of RAs is too small.
    ( ) Talent pool of RAs is too weak.
    ( ) Other (indicate below):
    _____
18. Using your best judgment, estimate the number of outside speakers who pass through your department in an average academic year.
    ( ) None   ( ) 16–20
    ( ) 1–5   ( ) 21–25
    ( ) 6–10   ( ) 26–30
    ( ) 11–15 ( ) More than 30

III. DEPARTMENTAL CHANGE

For the questions in this section, base your responses on your best judgment about how your department has changed over the past 10 years, that is, between roughly 1994 and 2004. Please respond to the questions as best as you are able, even if you were not a member of this department 10 years ago.

1. Compared to 10 years ago, how would you describe the constraints on your department's *overall operating resources*?
   ( ) Much more severe
   ( ) More severe
   ( ) About the same
   ( ) Less severe
   ( ) Much less severe
2. Compared to 10 years ago, how would you characterize the *equipment and research facilities* of your department?
   ( ) Much stronger

( ) Stronger
( ) About the same
( ) Weaker
( ) Much weaker

3. Compared to 10 years ago, how would you characterize constraints on *funding for graduate students* in your department?
   ( ) Much more severe
   ( ) More severe
   ( ) About the same
   ( ) Less severe
   ( ) Much less severe

4. Compared to 10 years ago, how would you characterize constraints on *funding for postdocs* in your department?
   ( ) Much more severe
   ( ) More severe
   ( ) About the same
   ( ) Less severe
   ( ) Much less severe

5. Compared to 10 years ago, how difficult is it for a typical faculty member to *publish in the top journals?*
   ( ) Much more difficult
   ( ) More difficult
   ( ) Just as difficult
   ( ) Easier
   ( ) Much easier

6. Compared to 10 years ago, how would you characterize constraints on *faculty travel funds?*
   ( ) Much more severe
   ( ) More severe
   ( ) About the same
   ( ) Less severe
   ( ) Much less severe

7. Compared to 10 years ago, how would you characterize the importance of faculty *external grants?*
   ( ) Much more important
   ( ) More important
   ( ) As important
   ( ) Less important
   ( ) Much less important

8. Compared to 10 years ago, how would you judge the competition for faculty *external grants*?
   ( ) Much more competitive
   ( ) More competitive
   ( ) As competitive
   ( ) Less competitive
   ( ) Much less competitive
9. Compared to 10 years ago, *earning tenure* in your department is . . . ?
   ( ) Much more difficult
   ( ) More difficult
   ( ) About as difficult
   ( ) Less difficult
   ( ) Much less difficult
10. Compared to 10 years ago, *earning promotion to full professor* in your department is...?
    ( ) Much more difficult
    ( ) More difficult
    ( ) About as difficult
    ( ) Less difficult
    ( ) Much less difficult
11. *Ten years ago*, how would you have judged the *morale of your department* in terms of the ability to conduct and complete scientific work?
    ( ) Very high
    ( ) High
    ( ) Average or Fair
    ( ) Low
    ( ) Very low
12. *Compared to 10 years ago*, how would you now describe the *morale of your department* in terms of the ability to conduct and complete scientific work?
    ( ) Much higher than 10 years ago
    ( ) Higher than 10 years ago
    ( ) About the same as 10 years ago
    ( ) Lower than 10 years ago
    ( ) Much lower than 10 years ago

IV. ADDITIONAL COMMENTS

Please make any additional comments you wish to make, including those that you feel would help inform any of your responses above.

Thank you for taking the time to complete this form.
Please return in the enclosed envelope.

APPENDIX G
Propositions Generated by the Study

The following propositions are derived from the present study and discussed in the corresponding sections of chapter five. They are presented in this appendix for summary and reference.

ON EXPECTATIONS AND THE RHYTHM OF CAREERS:

PROPOSITION 1: One observes notable reversals in outlook and identification with the career. In broad terms, elites enter mid-career highly satisfied only to end them with ambivalence. Communitarians enter mid-career highly dissatisfied and end them with serenity. In the middle, pluralists start on a "high," proceed to either a low or moderate level of satisfaction, and conclude on another "high."

PROPOSITION 2: By correspondence, attenuation in the latter-most phases of elites' careers is associated with lower levels of professional satisfaction and the development of an ambivalent work attitude. Withdrawal and separation in the latter-most phases of pluralists' and communitarians' respective careers is associated with higher levels of satisfaction (though not necessarily professional satisfaction), a positive work attitude among pluralists, and a detached attitude about work among communitarians.

ON ANOMIE AND ADAPTATION:

PROPOSITION 3A: At the end of their careers, elites customarily experience the phenomenon known as *anomie*. Communitarians and pluralists experience anomie also, but typically in much earlier phases of their careers, when it is possible for scientists in these worlds of science to realize that their career expectations cannot be realized.

PROPOSITION 3B: The incidence and longevity of anomie among elites is greatest because elites are exposed to the greatest potential for rewards.

PROPOSITION 3C: Anomie is more likely to arise when norms governing recognition are well developed, as in modern science, wherein the expectations to achieve recognition—transmitted organizationally through heightened university expectations in individual role performance—become pronounced.

PROPOSITION 4: Despite a high degree of control over their work, elites grow disillusioned with work and its rewards at the end of their careers. Despite a relatively lower degree of control over their work, communitarians and pluralists grow more satisfied at the end of their careers once they have left, or near to the time of leaving, work.

PROPOSITION 5A: Retreatism is most often enacted by communitarians and select pluralists because their universities offer a relative abundance of alternative rewards to research.

PROPOSITION 5B: The reason one observes retreatism in greater combination with ritualism among pluralists is because the press for scientific achievement is greater in the pluralist world.

PROPOSITION 5C: Elites most typically embrace ritualism because anomie characteristically hits them late in the career; their productive habits are so well formed and so routine that they do not shake free from them, instead continuing to produce without the knowledge that they will not be as recognized as desired.

PROPOSITION 5D: All scientists—communitarian, pluralist, and elite—attempt to be conformists. But they deviate, at varying times, and adopt alternative ways of adapting to their unfolding careers. Elites conform throughout the greatest portion of their careers.

PROPOSITION 5E: Rebellion as an adaptive process appears most likely in environments that allow the greatest degrees of decoupling from the institution of science, namely the communitarian, and to some extent, the pluralist worlds.

ON REFERENCE GROUPS AND SOCIAL CONTROL:

PROPOSITION 6: The manners by which scientists define success may be interpreted as a selection of reference groups that scientists adopt to calculate their achievements.

PROPOSITION 7: Some scientists realize both local and cosmopolitan identities over the course of a career; other scientists switch between them, gravitating toward one or the other in different career phases; and still other scientists embrace just one of these identities.

PROPOSITION 8: Many communitarians and some pluralists ended up bitter toward their employing organizations. And whereas they command and use specialized skills, and likely possess them well into their careers, the skills are not put to full use due to lack of opportunity and constraints that employing organizations place on professional careers.

PROPOSITION 9: The gap between reality and desire among elites is always great because the reference group they utilize to form self-judgments is always far removed and itself an embodiment of greatness. The gap between reality and future wants among communitarians closes because the reference group they come to utilize to form self-judgments is of their own making.

PROPOSITION 10: The ability of external reference groups to constrain individual behavior also establishes the conditions for comparatively continuous careers.

PROPOSITION 11: If turning points are occasions to revise identities, then organizational environments that spell fewer turning points in careers will allow for greater continuity in individual identity, a pattern of which elites are most representative.

ON SELECTION OF REFERENCE GROUPS:

PROPOSITION 12: Reference-group selection depends not exclusively on the operation of a professional reward system (i.e., the reward system of science) but also on organizational reward systems, those situated in the departments and universities that employ scientists.

PROPOSITION 13A: Plurality of legitimized choice can make it difficult to fail.

PROPOSITION 13B: The opportunity for reward in research will grow as a department is located toward the elite end of the spectrum; the opportunity for reward in teaching will grow as a department is located toward the communitarian end of the spectrum. But whether located toward either end or in the middle of the continuum, the pluralist department and institution will sanction, albeit in varying degrees, varieties of careers, since varieties of careers are alleged to be necessary for the organization's survival.

ON REJECTION OF REFERENCE GROUPS:

PROPOSITION 14: While organizationally suited reference groups control careers in their respective settings, not each and every scientist in a given setting conforms to the organizationally suited reference group. In any professional career, just as in any group, there is deviation.

PROPOSITION 15: The finding of sustained deviance further suggests the stricture with which the academic profession structures careers: once a deviant, always a deviant.

ON SOCIAL CONTROL OF THE LIFE COURSE:

PROPOSITION 16: The logic that emerges from the data strongly suggests that scientists develop the observed behavioral responses to their careers as their work, over time, is variously and disparately recognized by reward systems.

PROPOSITION 17: Noteworthy is the extent of discontent with the reward system of science: it is widespread, found across all the worlds of science and in all the cohorts of scientists.

PROPOSITION 18A: Widespread concern about the deficit of recognition provides further indication of the centrality that recognition plays in the construction, meaning, and evaluation of an academic career.

PROPOSITION 18B: The data convey a ubiquity of a viewpoint, suggestive of the character of the profession, that scientists regard their work as more significant than conveyed by the institutional recognition their work receives. Such a viewpoint establishes another ground for anomie, since individual beliefs about rewards deserved exceed the reality of rewards granted.

PROPOSITION 19A: Scientists cool out, but do so at different times, at different rates, and with different consequences, depending on the type of organizational environment in which the process occurs.

PROPOSITION 19B: In the case of science, much if not all of the cooling-out process takes place independently. The mark and cooler are one and the same.

PROPOSITION 19C: Structurally, it is difficult for members of a profession to discuss failure, and its many members are thus left partly paralyzed by it. The ineffectiveness of the cooling-out process in academia disallows the potential of completely "sending failure away," of effectively funneling it outside the social organization of work.

ON CAREERS IN OTHER ACADEMIC FIELDS:

PROPOSITION 20: Because elite, pluralist, and communitarian identities are organizationally based, it is logical to conclude that these worlds of work are inhabited by practitioners in all academic fields.

PROPOSITION 21: High expectations, and their evolution, situate careers not just in physics but in all academic fields. Marked by high expectations, particularly in the early phases of a career, across the three prototypes of American universities, careers may be seen to unfold in roughly similar ways.

PROPOSITION 22: Just as physics careers are organizationally and professionally bound, so, too, are careers in other academic fields. Both internal/organizational and external/professional systems of reward govern performance across fields.

PROPOSITION 23: The chance of disappointment is greatest in fields with the clearest collective minds, whereas the chance of disappointment is lowest in fields in disarray.

PROPOSITION 24: The high resource dependency and immutability of physics establish notable constraints on academic careers in that field and in fields like it.

PROPOSITION 25: If a field is highly resource dependent, then the function of cumulative advantage and disadvantage assumes particular significance.

PROPOSITION 26: One might expect practitioners in resource-independent and in mutable fields to offer less harsh judgments and more sanguine perspectives about their careers.

ON FUTURE COHORTS OF SCIENTISTS AND CONTEXTS OF SCIENCE:

PROPOSITION 27: If one were to find scientists (and other academics) who began their careers with more modest career expectations, it would tend to be in communitarian and in some pluralist institutions, which have since altered their missions and become more research-oriented, and then in the eldest cohorts of faculties in those institutions, who began with a less research-oriented emphasis.

PROPOSITION 28: There is far from definitive and emphatic enthusiasm for academia among those who have forged careers in its various institutions. The most positive responses come from elites, yet comparatively few academics will ever be employed by elite institutions.

PROPOSITION 29: While there may be a positive net change in research opportunity, this of course does not mean that there is congruency between the research expectations of individual scientists and those of their employing institutions. Where the former exceeds the latter, one may expect to witness the same cooling-out processes as found endemic throughout the present work.

PROPOSITION 30: Increased emphases on research will be accompanied by increased probabilities of anomie throughout the system of higher education.

NOTES

INTRODUCTION

1. United States Department of Labor, Bureau of Labor Statistics, *Technical Report 04–1797* (Washington, DC: U.S. Department of Labor, September 14, 2004), table 4.
2. Ibid.
3. Monthly hours computed: (40.45 hours/week × 52)/12 = 175.28. Annual hours computed in a 48-week year: (40.45 hours/week × 48) = 1,941.6.
4. Interview of William Faulkner conducted by Malcolm Cowley. In Malcolm Cowley (ed.), "William Faulkner," in *Writers at Work: The Paris Review Interviews* (New York: Viking, 1958). Quote appears on page 135. My thanks to Jeremy Reynolds for calling this passage to my attention and to Hugh Ruppersburg for helping to identify its source.
5. United States Department of Labor, Bureau of Labor Statistics. *Technical Report 04–1797*, table 2.
6. Waking hours working, computed: (24 hours/day × 5-day standard work week × 48-week year) − 1,982.4 = 3,777.6 hours awake. 1,941.6 working hours/3,777.6 hours awake = 51.4 percent.
7. Organization for Economic Cooperation and Development, "Clocking in and Clocking Out: Recent Trends in Working Hours,"*OECD Observer* (October 2004), 5, fig. 3. See also: International Labour Office, *Conditions of Work Digest* 14. 1995.
8. American Bar Foundation, *After the JD: First Results of a National Study of Legal Careers* (Chicago: American Bar Foundation, 2004), 33.
9. American Academy of Family Physicians, *Practice Profile Survey* (Leawood, KS: American Academy of Family Physicians, 2004) table 15. (Data include only active-member respondents of the American Academy of Family Physicians who are in office-based direct-patient care.)
10. Martin J. Finkelstein, Robert K. Seal, and Jack H. Schuster, *The New Academic Generation: A Profession in Transformation* (Baltimore: Johns Hopkins University Press, 1998) 71, table 25.
11. Scholars of higher education have considered where boundaries of a profes-

sion are drawn when referring to those people who are professors in universities and colleges (for example, Becker 1987; Becker and Trowler 1989; Clark 1987; Light 1974; Ruscio 1987). Professors simultaneously hold disciplinary, institutional, and academic statuses. Their disciplinary status refers to the specific area of specialization in which they work (e.g., physics and anthropology) and can be parsed further to refer to a specific specialty or mode of inquiry within a field (e.g., high-energy physics; physical anthropology; theorist; experimentalist). Their institutional status refers to the specific higher education organization in which they are employed. Their academic status refers to the broad occupational jurisdiction over which society has assigned a mandate and monopoly of service (only academics develop, organize, and transmit codified knowledge to successive generations through systematized instruction and research, and only academics confer degrees). Some scholars take the view that academia consists of fields and disciplines, each with their corresponding profession. Academia thus becomes a multitude of *professions*. The view has credence in the sense that fields differ in their cultures, histories, and styles. Nevertheless, academia remains their broader referent. To speak of an academic profession is not to deny its internal variation.

The same may be said for other professions. Pediatrics and cardiology may possess distinctive cultures, histories, and styles, and while pediatricians and cardiologists may be cognizant of such differences, they more broadly compose the medical profession. The medical profession, and not simply pediatricians or cardiologists, is socially assigned the mandate to heal the sick. Presbyterian pastors and Episcopal priests make up distinct denominational cultures, histories, and styles, but they more broadly compose the profession of ministry. The profession of ministry, and not simply Presbyterian pastors or Episcopal priests, is socially assigned the mandate to save souls. Divorce and corporate attorneys partake of distinct cultures, histories, and styles, but they more broadly compose the legal profession. The legal profession, and not simply divorce or corporate attorneys, is socially assigned the mandate to adjudicate disputes and negotiate exchanges between parties. Marines and army soldiers compose differing cultures, histories, and styles, but they more broadly compose the military profession. The military profession, and not simply marines or army soldiers, is socially assigned the mandate to defend a nation.

12. The U.S. Department of Education has conducted a series of national surveys of postsecondary faculty, the first in 1988, followed by 1993 and 2004 (results from the latest of which are unpublished and unavailable at this writing). These surveys do not track the same people over time, nor is the academic career the main object of the surveys' inquiry. Rather the surveys are intended primarily to collect cross-sectional demographic data on the conditions of faculty employment.

Logan Wilson is often credited for having written the first full-scale sociological study of the American academic profession, published in 1942 as *The Academic Man: A Study in the Sociology of a Profession*. He wrote a sequel to that work, published in 1979 as *American Academics: Then and Now*. Both of the books focus on academic structure, governance, and roles. The second book is a sequel in the sense of identifying several major changes in the structure of academe since his first study. Neither of the works draw on samples of academic practitioners, nor are academics followed over time to address continuity and change in their careers or the profession.

In her work on the psychology of scientists and scientific careers, Anne Roe appears to have come closest to conducting longitudinal inquiry on academics of any kind. Her first major work on scientists was published in 1952 as *The Making of a Scientist*. She reinterviewed the scientists in her original study roughly fifteen years later and collected select general data on changes in their productivity and work roles. But no full-scale follow-up study emerged on a par with the first one. Her selected longitudinal results were published in the article "Changes in Scientific Activities with Age: The Life of an Established Scientist Changes Little Over the Years—Unless He Goes into Administration" (Roe 1965) and in the article "Scientists Revisited," published in an internal Harvard University series on career development.

13. For a critical review of the Hughes oeuvre, see Chapoulie (1996) and Shaffir and Pawluck (2003).

14. For a discussion of the Chicago School of Sociology and the study of occupations, see Barley (1989).

15. For an anthology of longitudinal studies conducted in the social sciences, including description of their research designs, sample characteristics, and study parameters, see Young, Savoloa, and Phelps (1991).

16. Work by the social psychologist Melvin Kohn has explored, using quantitative analysis, the relationships between work and personality. His research in this vein has focused on tracing the effects of job conditions and occupational structure on social and psychological functioning (see Kohn 1969; Kohn and Schooler 1983; for related work, see Rosenberg 1957 and Spenner 1988). In the present work, I take the tack of employing qualitative analysis to explore the meanings that individuals derive about their locations in social structures of one profession.

CHAPTER ONE

1. At the time of the foundational study, only the first NRC assessment (Jones, Lindzey, and Coggeshall 1982) was available and was the one used for departmental sampling. A year after the fieldwork was completed, the second NRC assessment was published (Goldberger, Maher, and Flattau 1995). The rankings of the sampled departments between the two assessments is roughly similar. The greatest change was exhibited by the middle department, which improved in rank-

ing from 1982 to 1995, but which remained significantly below the top-ranked departments. A third assessment of graduate programs conducted by the NRC is scheduled for publication in 2008 but is not available at this writing.

2. Prior research has found that in the small number of cases of academics who move among academic jobs, mobility is best explained by specific achieved characteristics. Examining 274 job changes by physicists, chemists, mathematicians, and biologists between 1961 and 1975, Allison and Long (1987) found that the major determinants of prestige of the destination department were prestige of prior job, prestige of doctoral department, and number of articles published in the six years prior to a move. Crane (1970), focusing on the prestige of doctoral department, reported similar findings.

3. The top ten departments of physics in the United States as compiled from the assessment of research-doctorate programs conducted by the National Research Council and reported in Goldberger, Maher, and Flattau (1995). The departments include (eleven, due to a tie in ranking) from highest to lowest: Harvard, Princeton, M.I.T., University of California–Berkeley, Caltech, Cornell, Chicago, Illinois, Stanford, University of California–Santa Barbara, and Texas.

4. For related work concerned with timetables of careers and the salience of age in organizational environments, see Lashbrook (1996); Lawrence (1984; 1996); Roth (1963). For general theoretic work concerned with the temporal structuring of age in society, see Chudacoff (1989); Fry (2003); Kohli and Meyer (1986); Merton (1984); Settersten (1996; 1999, especially chapter 2; 2003).

5. Using event-history analysis on a sample of biochemists, Long, Allison, and McGinnis (1993) found that quantity of publications is more important than quality in predicting advancement in rank. "Time in rank and the number of publications in rank are the most important factors determining rates of promotion. There is little evidence that the quality of research, as indicated by citations to the articles or the standing of the journal in which the articles are published, affects promotion" (Long, Allison, and McGinnis 1993, 719). They also conclude that rates of rank advancement are lower for women and suggest that this disparity is explained by women being held to higher standards for promotion. "Everything else being equal, women are promoted to associate professor more slowly than men. . . . For promotions to full professor . . . . [b]eing in a more prestigious department has a significantly more negative effect on promotion for women. . . . At the rank of associate professor, women pay a greater price for being at prestigious departments, and while they receive a greater return for each publication, this provides an advantage only for the most-published female scientists" (Long, Allison, and McGinnis 1993, 720).

In other work, Cole and Cole (1973) argued that quality of publication is taken to be the more important criterion in the evaluation of scientists in elite departments whereas quantity of publication is taken to be the more important criterion in the evaluation of scientists in non-elite departments. They also found (1967) that quantity and quality of publication tends to be more highly correlated in elite departments.

6. The complete vitae from three scientists were unavailable. In two cases, all information except publication lists was obtained. One of these scientists had retired and no longer kept a complete vita, and a copy could not be found. In the other case, the complete publication list of the scientist was in the process of being converted into a different computer format, and despite a promise of its eventual mailing, was never received. In the third case, a retired scientist, no vita at all was obtained. Since this scientist had retired, of greatest importance was the record of publication up until and after the point of retirement (as opposed to change in position, rank, etc.). Vitae for each of these scientists had been collected at the time of the foundational study and thus presented their record of publication productivity up until that time. In all three cases, the publication records of these scientists established since the foundational study were compiled using the Science Citation Index. This effort thus produced a complete record of publication for all fifty-five members of the longitudinal sample.

7. For still additional examples of longitudinal interview work using samples of comparatively small sizes, see Young, Savola, and Phelps (1991).

CHAPTER TWO

1. Interview data are notated by number corresponding to each of the interviews. Interview data from the foundational and longitudinal fieldwork are separately designated. F follows the interview number corresponding to data from the foundational fieldwork (1F, etc.); L follows the interview number corresponding to data from the longitudinal fieldwork (1L, etc.). The interview numbers between the foundational and longitudinal sets of data are matched. For example, interviews 2F and 2L refer to data cited from interviewee no. 2, who is the identical interviewee between the foundational and longitudinal studies.

2. Merton (1973c) elaborates three other norms undergirding an ethos of science: communism (also referred to as *communalism*); disinterestedness; and organized skepticism. Communism stipulates that knowledge produced by scientists should be shared, not kept secret. Disinterestedness stipulates that scientists' conduct and performance of science should occur free from personal biases. Organized skepticism stipulates that scientists' judgments of the merits of contributions should occur only after all necessary evidence is on hand in order to render the most accurate judgments.

CHAPTER FOUR

1. Robert Weiss (2005) examines the multiple ways in which people experience retirement from varieties of work. Lorraine Dorfman (1997) examines professors'

experiential retirement patterns. Robert Clark and Brett Hammond (2001) examine retirement policy in U.S. higher education.

CHAPTER FIVE

1. Hagstrom's types are based on Merton's more general theoretical statement about social structure and anomie and types of individual adaptation (see Merton [1957] 1968a, esp. p. 193–211). Merton posits modes of individual adaptation in reference to cultural goals and institutionalized means of satisfying them, as summarized below.

Typology of Modes of Individual Adaptation

Modes of Adaptation	Cultural Goals	Institutionalized Means
I. Conformity	+	+
II. Innovation	+	−
III. Ritualism	−	+
IV. Retreatism	−	−
V. Rebellion	+/−	+/−

(*Source:* Merton [1957] 1968a, 194

2. For expanded discussion of deviance and illustration of cases in science and in the academic profession generally, see Braxton (1999) and LaFollette (1992).

3. Gouldner, too, builds his discussion on Merton, specifically his work on community influence (Merton [1957] 1968e).

4. Similar conclusions pertaining to career processes and the temporal conditions of identities in organizations apply to Glaser's generative work (Glaser 1963; 1964b; 1965).

5. Wilensky (1961) offered a discussion of the consequences of orderly versus "chaotic" work histories in terms of communal life and social participation. He observed that people with orderly careers, associated with high integration in the world of work, were more active participants of society: they belonged to more organizations, attended more civic and communal meetings, devoted more time to organizational activity, and developed greater attachment to local community than did those with disorderly work histories. Wilensky surmised that people following orderly careers, by virtue of their attendant characteristics, were strategic for social order. For Wilensky, *career* meant a work life in functionally related, hierarchically ordered sets of jobs. In the present case, all individuals have careers, but some are orderly, others less consistent. Among careers, both high and low integration in science is observed. Whereas Wilensky concluded that orderly careers are strategic

for social order, this study observes how orderly and disorderly careers may be necessary for the social order of science, especially in its various organizational contexts. Numerous academic organizations, especially communitarian and pluralist institutions, depend on disorderly careers. To fulfill their missions, varieties of careers are apparently necessary.

6. Survey research routinely reports academics as highly satisfied in their work (e.g., Blackburn and Lawrence 1995; Clark 1987; National Center for Education Statistics 1990; National Opinion Research Center 2000; Schuster and Finkelstein 2006), a discrepancy I have addressed in previous work (Hermanowicz 2003). Moreover, this research customarily finds high levels of satisfaction across institutional types, despite significant differences in work conditions among those types. Nor has prior survey research examined distinctions in satisfaction by career phase. The discrepancy in findings suggests that individuals, when completing questionnaires about job satisfaction, often provide normative responses that are consistent with the academic profession's prestige and with social norms that sanction the reporting of high satisfaction in work, especially in the work of the vaunted professions. When probed for details in the meaning and satisfaction of their work and various facets of their work roles in face-to-face interviews, individuals often offer more critical, candid, and finer-grained assessments. A lack of reported variation in response among a population that is itself highly variable on a number of characteristics should give pause and turn attention to the theoretically significant ways in which it likely bears notable permutations.

REFERENCES

Abbott, Andrew. 2004. *Methods of Discovery: Heuristics for the Social Sciences.* New York: Norton.
Allison, Paul D. and J. Scott Long. 1990. "Departmental Effects on Scientific Productivity." *American Sociological Review* 55:469–478.
———. 1987. "Interuniversity Mobility of Academic Scientists." *American Sociological Review* 52:643–652.
Allison, Paul D., J. Scott Long, and Tad K. Krauze. 1982. "Cumulative Advantage and Inequality in Science." *American Sociological Review* 47:615–625.
Allison, Paul D. and John A. Stewart. 1974. "Productivity Differences among Scientists: Evidence for Accumulative Advantage." *American Sociological Review* 39:596–606.
American Academy of Family Physicians. 2004. *Practice Profile Survey.* Leawood, KS: American Academy of Family Physicians.
American Bar Foundation. 2004. *After the JD: First Results of a National Study of Legal Careers.* Chicago: American Bar Foundation.
American Council on Education. 2001. *American Universities and Colleges.* New York: Walter de Gruyter.
———. 1992. *American Universities and Colleges.* New York: Walter de Gruyter.
American Institute of Physics. 2004. *2005 Graduate Programs in Physics, Astronomy, and Related Fields.* Melville, NY: American Institute of Physics.
———. 1994. "1993–94 Academic Workforce Report." *AIP Report: Education and Employment Statistics Division.* No. R-392.1 (December).
———. 1993. *Graduate Programs in Physics, Astronomy, and Related Fields, 1993–94.* New York: American Institute of Physics.
Anderson, Nels. 1923. *The Hobo.* Chicago: University of Chicago Press.
Atkinson, Robert. 1998. *The Life Story Interview.* Newbury Park, CA: Sage.
Baldwin, Roger G. and Robert T. Blackburn. 1981. "The Academic Career as a Developmental Process: Implications for Higher Education." *Journal of Higher Education* 52:598–614.
Ball, Donald W. 1976. "Failure in Sport." *American Sociological Review* 41:726–739.
Baltes, Paul B. and John R. Nesselroade. 1984. "Paradigm Lost and Paradigm Re-

gained: Critique of Dannefer's Portrayal of Life-Span Developmental Psychology." *American Sociological Review* 49:841–847.

Barley, Stephen R. 1989. "Careers, Identities, and Institutions: The Legacy of the Chicago School of Sociology." In Michael B. Arthur, Douglas T. Hall, and Barbara S. Lawrence (eds.), *Handbook of Career Theory*, 41–65. Cambridge: Cambridge University Press.

Barron's Educational Series. 2000. *Profiles of American Colleges, 2001*. netLibrary ebook.

———. 1996. *Profiles of American Colleges*. Hauppauge, NY: Barron's Educational Series.

Bayer, Alan E. and Jeffrey E. Dutton. 1977. "Career Age and Research-Professional Activities of Academic Scientists." *Journal of Higher Education* 48:259–282.

Becker, Howard S. 1982. *Art Worlds*. Berkeley: University of California Press.

———. 1964. "Personal Change in Adult Life." *Sociometry* 27:40–53.

———. 1951. *Role and Career Problems of the Chicago Public School Teacher*. Ph.D. diss., University of Chicago.

———. 1949. *The Professional Dance Musician in Chicago*. Master's thesis, University of Chicago.

Becker, Howard S., Blanche Geer, Everett Hughes, and Anselm L. Strauss. 1961. *Boys in White: Student Culture in Medical School*. Chicago: University of Chicago Press.

Becker, Tony. 1987. "The Disciplinary Shaping of the Profession." In Burton R. Clark (ed.), *The Academic Profession: National, Disciplinary, and Institutional Settings*, 271–303. Berkeley: University of California Press.

Becker, Tony and Paul R. Trowler. 1989. *Academic Tribes and Territories*. Buckingham: The Society for Research into Higher Education and Open University Press.

Ben-David, Joseph. 1972. "The Profession of Science and Its Powers." *Minerva* 10:362–383.

———. [1968] 1991. "Scientific Research and Economic Growth." In Gad Greudenthal (ed.), *Scientific Growth: Essays on the Social Organization and Ethos of Science*, 257–262. Berkeley: University of California Press.

———. 1963. "Professions in the Class System of Present-Day Societies." *Current Sociology* 12:247–330.

Bentley, Richard and Robert Blackburn. 1990. "Changes in Academic Research Performance Over Time: A Study of Institutional Accumulative Advantage." *Research in Higher Education* 31:327–353.

Bertaux, Daniel and Martin Kohli. 1984. "The Life Story Approach: A Continental View." *Annual Review of Sociology* 10:215–237.

Bjorklund, Diane. 1998. *Interpreting the Self*. Chicago: University of Chicago Press.

Blackburn, Robert T. and Janet H. Lawrence. 1995. *Faculty at Work: Motivation, Expectation, and Satisfaction*. Baltimore: Johns Hopkins University Press.

———. 1986. "Aging and the Quality of Faculty Job Performance." *Review of Educational Research* 23:265–290.

Blau, Peter and Otis Dudley Duncan. 1967. *The American Occupational Structure*. Glencoe, IL: Free Press.

Bledstein, Burton J. 1976. *The Culture of Professionalism: The Middle Class and the Development of Higher Education*. New York: W.W. Norton.

Braxton, John M. (ed.). 1999. *Perspectives on Scholarly Misconduct in the Sciences*. Columbus: Ohio State University Press.

———. 1983. "Department Colleagues and Individual Faculty Publication Productivity." *Review of Higher Education* 6:115–128.

Braxton, John M. and Lowell L. Hargens. 1996. "Variation among Academic Disciplines: Analytical Frameworks and Research." In John C. Smart (ed.), *Higher Education: Handbook of Theory and Research*, Vol. 11, 1–46. New York: Agathon.

Brint, Steven, Mark Riddle and Robert A. Hanneman. 2006. "Reference Sets, Identities, and Aspirations in a Complex Organizational Field: The Case of American Four-Year Colleges and Universities." *Sociology of Education* 79:229–252.

Burawoy, Michael. 2003. "Revisits: An Outline of a Theory of Reflexive Ethnography." *American Sociological Review* 68:645–679.

Cairns, Robert B., Lars R. Bergman, Jerome Kagan (eds.). 1998. *Methods and Models for Studying the Individual: Essays in Honor of Marian Radke-Yarrow*. Thousand Oaks, CA: Sage.

Carnegie Foundation for the Advancement of Teaching. 1994. *A Classification of Institutions of Higher Education, 1994 Edition*. Princeton, NJ: Carnegie Foundation for the Advancement of Teaching.

Carnegie Foundation for the Advancement of Teaching. 2000. *A Classification of Institutions of Higher Education, 2000 Edition*. Menlo Park, CA: Carnegie Foundation for the Advancement of Teaching.

Caspi, Avshalom, Glen H. Elder, Jr., and Ellen S. Herbener. 1990. "Childhood Personality and the Prediction of Life Course Patterns." In Lee Robbins and Michael Rutter (eds.), *Straight and Devious Pathways from Childhood to Adulthood*, 13–35. Cambridge: Cambridge University Press.

Chandrasekhar, S. 1987. *Truth and Beauty: Aesthetics and Motivations in Science*. Chicago: University of Chicago Press.

Chapoulie, Jean-Michel. 1996. "Everett Hughes and the Chicago Tradition." *Sociological Theory* 14:3–29.

Charmaz, Kathy. 2007. "Grounded Theory." In George Ritzer (ed.), *Encyclopedia of Sociology*, 2023–2027. Oxford: Blackwell.

———. 2001. "Qualitative Interviewing and Grounded Theory Analysis." In Jaber F. Gubrium and James A. Holstein (eds.), *Handbook of Interview Research: Context and Method*, 675–694. Thousand Oaks, CA: Sage.

———. 1990. "'Discovering' Chronic Illness: Using Grounded Theory." *Social Science and Medicine* 30:1161–1172.

Charmaz, Kathy and Richard G. Mitchell. 2001. "Grounded Theory in Ethnography." In Paul Atkinson, Amanda Coffey, Sara Delamont, John Lofland, and Lyn Lofland (eds.), *Handbook of Ethnography*, 160–174. Los Angeles: Sage.

Chudacoff, Howard P. 1989. *How Old Are You? Age Consciousness in American Culture.* Princeton, NJ: Princeton University Press.

Clair, Jeffrey, David Karp, and William C. Yoels. 1993. *Experiencing the Life Cycle.* Springfield, IL: Charles C. Thomas.

Clark, Burton R. 1987. *The Academic Life: Small Worlds, Different Worlds.* Princeton, NJ: Carnegie Foundation for the Advancement of Teaching.

Clark, Robert L. and P. Brett Hammond (eds.). 2001. *To Retire or Not? Retirement Policy and Practice in Higher Education.* Philadelphia: University of Pennsylvania Press.

Clausen, John A. 1998. "Life Reviews and Life Stories." In J. Giele and G. Elder, Jr. (eds.) *Methods of Life Course Research.* Thousand Oaks, CA: Sage.

———. 1986. *The Life Course.* Englewood Cliffs, NJ: Prentice Hall.

Clemente, Frank. 1973. "Early Career Determinants of Research Productivity." *American Journal of Sociology* 79:409–419.

Cohler, Bertram J. 1982. "Personal Narrative and Life Course." In Paul B. Baltes and Orville G. Brim, Jr. (eds.), *Life-Span Development and Behavior*, Vol. 4, 205–241. New York: Academic.

Cohler, Bertram J. and Andrew Hostetler. 2003. "Linking Life Course and Life Story: Social Change and the Narrative Study of Lives over Time." In Jeylan T. Mortimer and Michael J. Shanahan (eds.), *Handbook of the Life Course*, 555–576. New York: Kluwer Academic/Plenum Publishers.

Cole, Jonathon R. 1979. *Fair Science: Women in the Scientific Community.* New York: Columbia University Press.

Cole, Jonathan R. and Stephen Cole. 1973. *Social Stratification in Science.* Chicago: University of Chicago Press.

Cole, Jonathon R. and Harriet Zuckerman. 1987. "Marriage, Motherhood, and Research Performance in Science." *Scientific American* 256:119–125.

Cole, Stephen. 1992. *Making Science: Between Nature and Society.* Cambridge, MA: Harvard University Press.

———. 1983. "The Hierarchy of the Sciences?" *American Journal of Sociology* 89:111–139.

———. 1979. "Age and Scientific Performance." *American Journal of Sociology* 84:958–977.

———. 1970. "Professional Standing and the Reception of Scientific Discoveries." *American Journal of Sociology* 76:286–306.

Cole, Stephen and Jonathan R. Cole. 1967. "Scientific Output and Recognition: A

Study in the Operation of the Reward System in Science." *American Sociological Review* 32:377–390.

Cole, Stephen, Leonard Rubin, and Jonathan Cole. 1978. *Peer Review in the National Science Foundation: Phase I.* Washington, DC: National Academy of Sciences.

Cole, Stephen, Gary Simon, and Jonathan R. Cole. 1988. "Do Journal Rejection Rates Index Consensus?" *American Sociological Review* 53:152–156.

Cowley, Malcolm. 1958. "William Faulkner." In Malcolm Cowley (ed.) *Writers at Work: The Paris Review Interviews,* 117–141. New York: Viking.

Crane, Diana. 1970. "The Academic Marketplace Revisited: A Study of Faculty Mobility Using the Cartter Ratings." *American Journal of Sociology* 75:953–964.

———. 1965. "Scientists at Major and Minor Universities: A Study of Productivity and Recognition." *American Sociological Review* 30:600–714.

Cressey, Paul. 1932. *The Taxi-Dance Hall.* Chicago: University of Chicago Press.

Dannefer, Dale. 1987. "Aging as Intracohort Differentiation: Accentuation, the Matthew Effect, and the Life Course." *Sociological Forum* 2:211–236.

———. 1984a. "Adult Development and Social Theory: A Paradigmatic Reappraisal." *American Sociological Review* 49:100–116.

———. 1984b. "The Role of the Social in Life-Span Developmental Psychology, Past and Future: Rejoinder to Baltes and Nesselroade." *American Sociological Review* 49:847–850.

Davis, Fred. 1951. *The Community Newspaper in Metropolitan Chicago.* Master's thesis, University of Chicago.

Denzin, Norman. 1989. *Interpretive Biography.* Newbury Park, CA: Sage.

DiPrete, Thomas A. and Gregory M. Eirich. 2006. "Cumulative Advantage as a Mechanism for Inequality: A Review of Theoretical and Empirical Developments." *Annual Review of Sociology* 32:271–297.

Donovan, Francis. 1929. *The Saleslady.* Chicago: University of Chicago Press.

Dorfman, Lorraine T. 1997. *The Sun Still Shone: Professors Talk About Retirement.* Iowa City: University of Iowa Press.

Dubin, Robert. 1992. *Central Life Interests: Creative Individualism in a Complex World.* New Brunswick, NJ: Transaction.

Dunham, Edgar Alden. 1969. *Colleges of the Forgotten Americans: A Profile of State Colleges and Regional Universities.* New York: McGraw-Hill.

Durkheim, Emile. [1915] 1965. *The Elementary Forms of the Religious Life.* New York: Free Press.

———. [1897] 1951. *Suicide.* Translated by John A. Spalding and George Simpson. New York: Free Press.

———. [1895] 1982. *The Rules of Sociological Method.* Edited with an introduction by Steven Lukes. Translated by W.D. Halls. New York: Free Press.

Dyson, Freeman. 1979. *Disturbing the Universe.* New York: Basic.

Eiduson, Bernice T. 1962. *Scientists: Their Psychological World.* New York: Basic.

Eisenhart, Margaret A. and Elizabeth Finkel. 1998. *Women's Science: Learning and Succeeding from the Margins*. Chicago: University of Chicago Press.

Elder, Glen H. Jr. 1998. "The Life Course and Human Development." In Richard M. Lerner (ed.), *Handbook of Child Psychology*. Vol. 1, *Theoretical Models of Human Development*, 939–991. New York: Wiley.

———. 1985. "Perspectives on the Life Course." In Glen H. Elder Jr. (ed.), *Life Course Dynamics: Trajectories and Transitions, 1968–1980*. Ithaca: Cornell University Press.

———. 1981. "History and the Life Course." In Daniel Bertaux (ed.), *Biography and Society: The Life History Approach in the Social Sciences*, 77–115. Newbury Park, CA: Sage.

———. 1975. "Age Differentiation and the Life Course." *Annual Review of Sociology* 1:165–190.

———. 1974. *Children of the Great Depression: Social Change in Life Experience*. Chicago: University of Chicago Press.

Erikson, Erik H. 1950. *Childhood and Society*. New York: Norton.

Faulkner, Robert R. 1985. *Hollywood Studio Musicians: Their Work and Careers in the Recording Industry*. New York: University Press of America.

———. 1974. "Coming of Age in Organizations: A Comparative Study of Career Contingencies and Adult Socialization." *Sociology of Work and Occupations* 1:131–173.

Finkelstein, Martin J., Robert K. Seal, and Jack H. Schuster. 1998. *The New Academic Generation: A Profession in Transformation*. Baltimore: Johns Hopkins University Press.

Finnegan, D.E. and Z.F. Gamson. 1996. "Disciplinary Adaptations to Research Culture in Comprehensive Institutions." In D.E. Finnegan, D.E. Webster, and Z.F. Gamson (eds.), *Faculty and Faculty Issues in Colleges and Universities*, 2nd ed., 476–498. ASHE Reader Series. Needham Heights, MA: Simon and Schuster Custom Publishing.

Fox, Mary Frank. 2005. "Gender, Family Characteristics, and Publication Productivity among Scientists." *Social Studies of Science* 35:131–150.

———. 1995. "Women and Scientific Careers." In S. Jasanoff, G.E. Markle, J.C. Petersen, and T. J. Pinch (eds.), *Handbook of Science and Technology Studies*, 205–223. Thousand Oaks, CA: Sage.

———. 1985. "Publication, Performance, and Reward in Science and Scholarship." In J. Smart (ed.), *Higher Education: Handbook of Theory and Research*, 255–282. New York: Agathon.

Friedson, Elliot. 1975. *Doctoring Together: A Study of Professional Social Control*. Chicago: University of Chicago Press.

———. 1970. *Profession of Medicine: A Study of the Sociology of Applied Knowledge*. Chicago: University of Chicago Press.

———. 1960. "Client Control and Medical Practice." *American Journal of Sociology* 65:374–382.

Fry, Christine L. 2003. "The Life Course as a Cultural Construct." In Richard A. Settersten (ed.), *Invitation to the Life Course: Toward New Understandings of Later Life*, 269–294. Amityville, NY: Baywood.

Gaston, Jerry. 1978. *The Reward System in British and American Science*. New York: Wiley.

Geertz, Clifford. 1983. *Local Knowledge*. New York: Basic.

———. 1973. *The Interpretation of Cultures*. New York: Basic.

Giele, Janet Z. and Glen H. Elder, Jr. (eds.). 1998. *Methods of Life Course Research: Qualitative and Quantitative Approaches*. Thousand Oaks, CA: Sage.

Glaser, Barney G. 1965. "'Differential Association' and the Institutional Motivation of Scientists." *Administrative Science Quarterly* 10:82–97.

———. 1964a. "Comparative Failure in Science." *Science* 143:1012–1014.

———. 1964b. *Organizational Scientists: Their Professional Careers*. Indianapolis, IN: Bobbs-Merrill.

———. 1963. "Variations in the Importance of Recognition in Scientists' Careers." *Social Problems* 10:268–276.

Glaser, Barney and Anselm L. Strauss. 1971. *Status Passage*. Chicago: Aldine.

Glashow, Sheldon L. 1991. *The Charm of Physics*. New York: Simon and Schuster.

Goffman, Erving. 1961. "The Moral Career of the Mental Patient." In Erving Goffman, *Asylums: Essays on the Social Situation of Mental Patients and other Inmates*, 127–169. Garden City, NY: Anchor. Article first published in 1959.

———. 1952. "On Cooling the Mark Out." *Psychiatry* 15:451–463.

Goldberger, Marvin L., Brendan A. Maher, and Pamela Ebert Flattau (eds.). 1995. *Research-Doctorate Programs in the United States: Continuity and Change*. Washington, DC: National Academy Press.

Goldner, Fred H. 1965. "Demotion in Industrial Management." *American Sociological Review* 30:714–724.

Goldner, Fred H. and R. R. Ritti. 1967. "Professionalization as Career Immobility." *American Journal of Sociology* 72:489–502.

Goode, William J. 1978. *The Celebration of Heroes: Prestige as a Control System*. Berkeley: University of California Press.

———. 1967. "The Protection of the Inept." *American Sociological Review* 32:5–19.

Gouldner, Alvin W. 1957–58. "Cosmopolitans and Locals: Toward an Analysis of Latent Social Roles I and II." *Administrative Science Quarterly* 2:281–306.

Gustin, Bernard H. 1973. "Charisma, Recognition, and the Motivation of Scientists." *American Journal of Sociology* 78:1119–1134.

Habenstein, Robert. 1954. *The Career of the Funeral Director*. Ph.D. diss., University of Chicago.

Haber, Samuel. 1991. *The Quest for Authority and Honor in the American Professions, 1750–1900*. Chicago: University of Chicago Press.

Hagstrom, Warren O. 1965. *The Scientific Community*. New York: Basic.

———. 1964. "Anomy in Scientific Communities." *Social Problems* 12:196–195.

Hargens, Lowell L. 1988. "Scholarly Consensus and Journal Rejection Rates." *American Sociological Review* 53:139–151.

———. 1978. "Relations between Work Habits, Research Technologies, and Eminence in Science." *Sociology of Work and Occupations* 5:97–112.

———. 1975. *Patterns of Scientific Research: A Comparative Analysis of Research in Three Scientific Fields*. Washington, DC: American Sociological Association.

Hargens, Lowell L. and Warren O. Hagstrom. 1967. "Sponsored and Contest Mobility of American Academic Scientists." *Sociology of Education* 40:24–38.

Hargens, Lowell L. and Lisa Kelly-Wilson. 1994. "Determinants of Disciplinary Discontent." *Social Forces* 72:1177–1195.

Harper, Douglas and Helene M. Lawson (eds.). 2003. *The Cultural Study of Work*. Lanham, MD: Rowman and Littlefield.

Hayner, Norman. 1936. *Hotel Life*. Durham: University of North Carolina Press.

Henderson, B.B. and W.D. Kane. 1991. "Caught in the Middle: Faculty and Institutional Status and Quality in State Comprehensive Universities." *Higher Education* 22:339–350.

Hermanowicz, Joseph C. 2005. "Classifying Universities and Their Departments: A Social World Perspective." *Journal of Higher Education* 76:26–55.

———. 2003. "Scientists and Satisfaction." *Social Studies of Science* 33:45–73.

———. 2002. "In the Shadows of Giants: Identity and Institution Building in the American Academic Profession." In Richard A. Settersten, Jr. and Timothy J. Owens (eds.), *Advances in Life Course Research, Vol. 7, New Frontiers in Socialization*, 133–162. London: Elsevier.

———. 1998. *The Stars Are Not Enough: Scientists—Their Passions and Professions*. Chicago: University of Chicago Press.

Hernstein, Richard J. and Charles Murray. 1994. *The Bell Curve: Intelligence and Class Structure in American Life*. New York: Free Press.

Hiles, Sara Shipley. January 23, 2006. "Young Scientists Hit the Hardest as U.S. Funding Falls." *Boston Globe*, Science Section.

Hughes, Everett C. 1997. "Careers." *Qualitative Sociology* 20(3):389–397.

———. 1994. *Everett C. Hughes: On Work, Race, and the Sociological Imagination*. Edited and with an Introduction by Lewis A. Coser. Chicago: University of Chicago Press.

———. 1971. *The Sociological Eye: Selected Papers*. New Brunswick, NJ: Transaction.

———. 1958a. *Men and Their Work*. Glencoe, IL: Free Press.

———. 1958b. "Cycles, Turning Points, and Careers." In Everett C. Hughes, *Men and Their Work*, 11–22. Glencoe, IL: Free Press.

———. 1937. "Institutional Office and the Person." *American Journal of Sociol-*

ogy 43:404–413. Reprinted in Everett C. Hughes. 1971. *The Sociological Eye: Selected Papers*, 132–140. New Brunswick, NJ: Transaction.

———. 1928. *A Study of a Secular Institution: The Chicago Real Estate Board*. Ph.D. diss., University of Chicago.

International Labour Office. 1995. *Conditions of Work Digest*. Vol. 14. Geneva: International Labour Office.

Jones, Lyle V., Gardner Lindzey, and Porter E. Coggeshall (eds.). 1982. *An Assessment of Research-Doctorate Programs in the United States: Mathematical and Physical Sciences*. Washington, DC: National Academy Press.

Kalleberg, Arne L. 1977. "Work Values and Job Rewards: A Theory of Job Satisfaction." *American Sociological Review* 42:124–143.

Kalleberg, Arne L. and Karyn A. Loscocco. 1983. "Aging, Values, and Rewards: Explaining Age Differences in Job Satisfaction." *American Sociological Review* 48:78–90.

Kohli, Martin and John W. Meyer. 1986. "Social Structure and Social Construction of Life Stages." *Human Development* 29:145–180.

Kohn, Melvin L. 1976. "Occupational Structure and Alienation." *American Journal of Sociology* 82:111–130.

———. 1969. *Class and Conformity: A Study in Values*. Homewood, IL: Dorsey Press.

Kohn, Melvin L. and Carmi Schooler. 1983. *Work and Personality: An Inquiry into the Impact of Social Stratification*. Norwood, NJ: Ablex.

LaFollette, Marcel C. 1992. *Stealing into Print: Fraud, Plagiarism, and Misconduct in Scientific Publishing*. Berkeley: University of California Press.

Lashbrook, Jeff. 1996. "Promotional Timetables: An Exploratory Investigation of Age Norms for Promotional Expectations and Their Association with Job Well-Being." *Gerontologist* 36:189–198.

Laub, John H. and Robert J. Sampson. 2003. *Shared Beginnings, Divergent Lives: Delinquent Boys to Age 70*. Cambridge, MA: Harvard University Press.

Lawrence, Barbara S. 2006. "Organizational Reference Groups: A Missing Perspective on Social Context." *Organization Science* 17:80–100.

———. 1996. "Organizational Age Norms: Why Is It So Hard to Know When You See One?" *Gerontologist* 36:209–220.

———. 1984. "Age Grading: The Implicit Organizational Timetable." *Journal of Occupational Behaviour* 5:23–35.

Lawrence, Janet H. and Robert T. Blackburn. 1985. "Faculty Careers: Maturation, Demographic, and Historical Effects." *Research in Higher Education* 22:135–154.

Lehman, Harvey C. 1958. *Age and Achievement*. Princeton, NJ: Princeton University Press.

Levinson, Daniel J. 1996. *The Seasons of a Woman's Life*. New York: Alfred A. Knopf.

———. 1978. *The Seasons of a Man's Life*. New York: Ballantine.

Light, Donald, Jr. 1974. "The Structure of the Academic Professions." *Sociology of Education* 47:2–28.

Linde, Charlotte. 1993. *Life Stories: The Creation of Coherence.* New York: Oxford University Press.

Long, J. Scott. 1978. "Productivity and Academic Position in the Scientific Career." *American Sociological Review* 43:889–908.

Long, J. Scott, Paul D. Allison, and Robert McGinnis. 1993. "Rank Advancement in Academic Careers: Sex Differences and the Effects of Productivity." *American Sociological Review* 58:703–722.

———. 1979. "Entrance into the Academic Career." *American Sociological Review* 44:816–830.

Long, J. Scott and Mary Frank Fox. 1995. "Scientific Careers: Universalism and Particularism." *Annual Review of Sociology* 21:45–71.

Long, J. Scott and Robert McGinnis. 1981. "Organizational Context and Scientific Productivity." *American Sociological Review* 46:422–442.

Martin, Joanne. 1982. "Stories and Scripts in Organizational Settings." In Albert H. Hastorf and Alice M. Isen (eds.), *Cognitive Social Psychology,* 255–305. New York: Elsevier.

McClelland, David C. 1961. *The Achieving Society.* New York: Free Press.

McCormmach, Russell. 1982. *Night Thoughts of a Classical Physicist.* Cambridge, MA: Harvard University Press.

Merton, Robert K. 1984. "Socially Expected Durations: A Case Study of Concept Formation in Sociology." In W. W. Powell and R. Robbins (eds.), *Conflict and Consensus: A Festschrift for Lewis A. Coser,* 265–281. New York: Free Press.

———. 1977. *The Sociology of Science: An Episodic Memoir.* Carbondale, IL: Southern Illinois University Press.

———. 1976. *Sociological Ambivalence and Other Essays.* New York: Free Press.

———. 1973a. "Priorities in Scientific Discovery." In *The Sociology of Science: Theoretical and Empirical Investigations,* 286–324. Edited and with an Introduction by Norman W. Storer. Chicago: University of Chicago Press. Article first published in 1957.

———. 1973b. "The Matthew Effect in Science." In *The Sociology of Science: Theoretical and Empirical Investigations,* 439–459. Edited and with an Introduction by Norman W. Storer. Chicago: University of Chicago Press. Article originally published in 1968.

———. 1973c. "The Normative Structure of Science." In *The Sociology of Science: Theoretical and Empirical Investigations,* 267–278. Edited and with an Introduction by Norman W. Storer. Chicago: University of Chicago Press. Article originally published in 1942.

———. [1957] 1968a. "Social Structure and Anomie." In Robert K. Merton, *Social Theory and Social Structure,* 185–214. New York: Free Press.

———. [1957] 1968b. "Continuities in the Theory of Social Structure and Anomie." In Robert K. Merton, *Social Theory and Social Structure,* 215–248. New York: Free Press.

---. [1957] 1968c (with Alice S. Rossi). "Contributions to the Theory of Reference Group Behavior." In Robert K. Merton, *Social Theory and Social Structure*, 279–334. New York: Free Press.

---. [1957] 1968d. "Continuities in the Theory of Reference Groups and Social Structure." In Robert K. Merton, *Social Theory and Social Structure*, 335–440. New York: Free Press.

---. [1957] 1968e. "Patterns of Influence: Locals and Cosmopolitan Influentials." In Robert K. Merton, *Social Theory and Social Structure*, 441–474. New York: Free Press.

Merton, Robert K. and Harriet Zuckerman. 1973. "Age, Aging, and Age Structure in Science." In Robert K. Merton (ed.), *The Sociology of Science: Theoretical and Empirical Investigations*, 497–559. Chicago: University of Chicago Press. Article first published in 1972.

Monastersky, Richard. 2007. "The Real Science Crisis: Bleak Prospects for Young Researchers." *The Chronicle of Higher Education*, vol. LIV, no. 4 (September 21), 1.

Morphew, Christopher C. 2002. "'A Rose by Any Other Name': Which Colleges Became Universities." *Review of Higher Education* 25:207–223.

Morse, Janice M. 1994. "Designing Funded Qualitative Research." In Norman K. Denzin and Yvonna S. Lincoln (eds.) *Handbook of Qualitative Research*, 220–235. Thousand Oaks, CA: Sage.

Mortimer, Jeylan T. and Roberta G. Simmons. 1978. "Adult Socialization." *Annual Review of Sociology* 4:421–454.

Morus, Iwan Rhys. 2005. *When Physics Became King*. Chicago: University of Chicago Press.

National Center for Education Statistics. 1997. *Instructional Faculty and Staff in Higher Education Institutions: Fall 1987 and Fall 1992*. Washington, DC: U.S. Department of Education.

---. 1996. *Institutional Policies and Practices Regarding Faculty in Higher Education*. Washington, DC: U.S. Department of Education.

---. 1994. *Faculty and Instructional Staff: Who Are They and What Do They Do?* Washington, DC: U.S. Department of Education.

---. 1990. *Faculty in Higher Education Institutions, 1988*. Washington, DC: U.S. Department of Education.

National Opinion Research Center. 2000. *National Opinion Research Center, The American Faculty Poll*. Chicago: National Opinion Research Center.

National Research Council. 2001. *From Scarcity to Visibility: Gender Differences in the Careers of Doctoral Scientists and Engineers*. Washington, DC: National Academy Press.

National Science Board. 2004. *Science and Engineering Indicators, 2004*. Washington, DC: U.S. Government Printing Office.

Neave, Guy. 1979. "Academic Drift: Some Views from Europe." *Studies in Higher Education* 4:143–159.

Neugarten, Bernice L. 1996. *The Meanings of Age: Selected Papers of Bernice L. Neugarten.* Edited by Dail A. Neugarten. Chicago: University of Chicago Press.

———. 1979. "Time, Age, and the Life Cycle." *American Journal of Psychiatry* 136:887–894.

——— (ed). 1968. *Middle Age and Aging: A Reader in Social Psychology.* Chicago: University of Chicago Press.

Neugarten, Bernice L. and Nancy Datan. 1973. "Sociological Perspectives on the Life Cycle." In Paul B. Baltes and K. Warner Schaie (eds.), *Life-Span Development Psychology: Personality and Socialization,* 53–79. New York: Academic Press.

Neugarten, Bernice L., Joan W. Moore, and John C. Lowe. 1965. "Age Norms, Age Constraints, and Adult Socialization." *American Journal of Sociology* 70:710–717.

Organization for Economic Cooperation and Development. OECD Observer, October, 2004. "Clocking in and Clocking Out: Recent Trends in Working Hours." Paris: Organization for Economic Cooperation and Development.

Parkes, Colin M. and Robert S. Weiss. 1983. *Recovery from Bereavement.* New York: Basic.

Parsons, Talcott and Gerald M. Platt. 1973. *The American University.* Cambridge, MA: Harvard University Press.

Paul, Charles B. 1980. *Science and Immortality: The Eloges of the Paris Academy of Sciences, 1699–1791.* Berkeley: University of California Press.

Pelz, Donald C. and Frank M. Andrews. 1966. *Scientists in Organizations: Productive Climates for Research and Development.* New York: Wiley.

Preston, Anne E. 2004. *Leaving Science: Occupational Exit from Scientific Careers.* New York: Russell Sage Foundation.

Ragin, Charles C. 1987. *The Comparative Method: Moving Beyond Qualitative and Quantitative Strategies.* Berkeley: University of California Press.

Reskin, Barbara. 1979. "Age and Scientific Productivity: A Critical Review." In Michael S. McPherson (ed.), *The Demand for New Faculty in Science and Engineering.* Washington, DC: National Academy of Sciences.

———. 1978a. "Academic Sponsorship and Scientists' Careers." *Sociology of Education* 52:129–146.

———. 1978b. "Scientific Productivity, Sex, and Location in the Institution of Science." *American Journal of Sociology* 83:1235–1243.

———. 1977. "Scientific Productivity and the Reward Structure of Science." *American Sociological Review* 42:491–504.

———. 1976. "Sex Differences in Status Attainment in Science: The Case of Post-Doctoral Fellowships." *American Sociological Review* 41:597–613.

Rhoades, Gary. 1998. *Managed Professionals: Unionized Faculty and Restructuring Academic Labor.* Albany: State University of New York Press.

Riley, Matilda White, Anne Foner, and Joan Waring. 1988. "Sociology of Age." In Neil J. Smelser (ed.), *Handbook of Sociology,* 243–290. Newbury Park: CA: Sage.

Roe, Anne. 1965. "Changes in Scientific Activities with Age: The Life of an Established Scientist Changes Little Over the Years—Unless He Goes into Administration." *Science* 150:313–318.

———. 1952. *The Making of a Scientist*. New York: Dodd, Mead.

Rosenberg, Morris. 1957. *Occupations and Values*. New York: Free Press.

Roth, Julius A. 1963. *Timetables: Structuring the Passage of Time in Hospital Treatment and Other Careers*. Indianapolis, IN: Bobbs-Merrill.

Rubin, David C. (ed.). 1986. *Autobiographical Memory*. Cambridge: Cambridge University Press.

Rubin, Herbert J. and Irene S. Rubin. 1995. *Qualitative Interviewing: The Art of Hearing Data*. Thousand Oaks, CA: Sage.

Ruch, Richard S. 2001. *Higher Ed, Inc.: The Rise of the For-Profit University*. Baltimore: Johns Hopkins University Press.

Ruscio, Kenneth P. 1987. "Many Sectors, Many Professions." In Burton R. Clark (ed.), *The Academic Profession: National, Disciplinary, and Institutional Settings*, 331–368. Berkeley: University of California Press.

Saldana, Johnny. 2003. *Longitudinal Qualitative Research*. Walnut Creek, CA: AltaMira.

Schuster, Jack H. and Martin J. Finkelstein. 2006. *The American Faculty: The Restructuring of Academic Work and Careers*. Baltimore: Johns Hopkins University Press.

Settersten, Richard A., Jr. 2003. "Age Structuring and the Rhythm of the Life Course." In Jeylan T. Mortimer and Michael J. Shanahan (eds.), *Handbook of the Life Course*, 81–98. New York: Kluwer.

———. 1999. *Lives in Time and Place: The Problems and Promises of Developmental Science*. Amityville, NY: Baywood.

———. 1996. "What's the Latest? Cultural Age Deadlines for Educational and Work Transitions." *Gerontologist* 36:602–613.

Shaffir, William and Dorothy Pawluch. 2003. "Occupations and Professions." In Larry T. Reynolds and Nancy J. Herman-Kinney (eds.), *Handbook of Symbolic Interactionism*, 893–913. New York: AltaMira.

Shaw, Clifford. 1930. *The Jack-Roller: A Delinquent Boy's Own Story*. Chicago: University of Chicago Press.

Sheehy, Gail. 1976. *Passages: Predictable Crises of Adult Life*. New York: Bantam.

Shibutani, Tamotsu. 1962. "Reference Groups and Social Control." In Arnold Rose (ed.), *Human Behavior and Social Processes*, 128–145. Boston: Houghton Mifflin.

———. 1955. "Reference Groups as Perspectives." *American Journal of Sociology* 60:562–569.

Simon, Rita James. 1974. "The Work Habits of Eminent Scholars." *Sociology of Work and Occupations* 1:327–335.

Sonnert, Gerhard and Gerald Holton. 1995. *Gender Differences in Science Careers: The Project Access Study*. New Brunswick, NJ: Rutgers University Press.

Spenner, Kenneth I. 1988. "Social Stratification, Work, and Personality." *Annual Review of Sociology* 14:69–97.

Stebbins, Richard A. 1970. "Career: The Subjective Approach." *Sociological Quarterly* 11:32–49.

Stephan, Paula and Sharon Levin. 1992. *Striking the Mother Lode in Science: The Importance of Age, Place, and Time.* Oxford: Oxford University Press.

Strauss, Anselm L. and Juliet Corbin. 1994. "Grounded Theory Methodology: An Overview." In Norman K. Denzin and Yvonna S. Lincoln (eds.), *Handbook of Qualitative Research*, 273–285. Thousand Oaks, CA: Sage.

Strauss, Anselm L. and Lee Rainwater. 1962. *The Professional Scientist: A Study of American Chemists.* Chicago: Aldine.

Thomas, W.I. 1923. *The Unadjusted Girl: With Cases and Standpoint for Behavioral Analysis.* Boston: Little, Brown.

Trollope, Anthony. [1883] 1950. *An Autobiography.* Oxford: Oxford University Press.

United States Department of Labor. September 14, 2004. Bureau of Labor Statistics, *Technical Report* 04–1797. Washington, DC: U.S. Department of Labor.

Vaillant, George E. 1977. *Adaptation to Life.* Cambridge, MA: Harvard University Press.

Van Maanen, John (ed.). 1977. *Organizational Careers: Some New Perspectives.* New York: John Wiley and Sons.

Wanner, Richard A., Lionel S. Lewis, and David I. Gregorio. 1981. "Research Productivity in Academia: A Comparative Study of the Sciences, Social Sciences and Humanities." *Sociology of Education* 54:238–253.

Weiss, Robert S. 2005. *The Experience of Retirement.* Ithaca: Cornell University Press.

———. 1994. *Learning from Strangers: The Art and Method of Qualitative Interview Studies.* New York: Free Press.

Wells, Edward and Sheldon Stryker. 1988. "Stability and Change in Self over the Life Course." In Paul B. Baltes and David L. Featherman (eds.), *Life-Span Development and Behavior*, Vol. 8, 191–229. Hillsdale, NJ: Lawrence Erlbaum.

White, Robert. 1966. *Lives in Progress: A Study of the Natural Growth of Personality.* New York: Holt, Rinehart, and Winston.

Wilensky, Harold L. 1961. "Orderly Careers and Social Participation: The Impact of Work History on Social Integration in the Middle Mass." *American Sociological Review* 26:521–539.

Wilson, Logan. 1979. *American Academics: Then and Now.* New York: Oxford University Press.

———. 1942. *The American Academic Profession: A Study in the Sociology of a Profession.* New York: Oxford University Press.

Xie, Yu and Kimberlee A. Shauman. 2003. *Women in Science: Career Processes and Outcomes.* Cambridge, MA: Harvard University Press.

Young, Copeland H., Kristen L. Savola, and Erin Phelps. 1991. *Inventory of Longitudinal Studies in the Social Sciences.* Newbury Park, CA: Sage.

Zuckerman, Harriet. 1998. "Accumulation of Advantage and Disadvantage: The Theory and Its Intellectual Biography." In Carlo Mongardini and Simonetta Tabboni (eds.), *Robert K. Merton and Contemporary Sociology*, 139–161. New Brunswick, NJ: Transaction.

———. 1992. "The Proliferation of Prizes: Nobel Complements and Nobel Surrogates in the Reward System of Science." *Theoretical Medicine* 13:217–231.

———. 1988. "Sociology of Science." In Neil J. Smelser (ed.), *Handbook of Sociology*, 511–574. Newbury Park, CA: Sage.

———. 1977. *Scientific Elite: Nobel Laureates in the United States*. New York: Free Press.

———. 1970. "Stratification in American Science." *Sociological Inquiry* 40:235–257.

Zuckerman, Harriet and Jonathan R. Cole. 1975. "Women in American Science." *Minerva* 13:82–102.

Zuckerman, Harriet, Jonathon R. Cole, and John T. Bruer (eds.). 1991. *The Outer Circle: Women in the Scientific Community*. New York: Norton.

Zuckerman, Harriet and Robert K. Merton. 1971. "Patterns of Evaluation in Science: Institutionalization, Structure and Functions of the Referee System." *Minerva* 9:66–100.

INDEX

Abbott, Andrew, 53, 54
Age and aging, 3, 31, 47–48, 76, 138; and achievement, 246–247; continuities and changes in, 3–4, 19–20; meanings of, 9–11, 298n4; scientists distributions of, 43–45, 77–78, 139–141, 179–181
Allison, Paul, 13, 95, 246, 247, 298n2, 298n5
American Academy of Arts and Sciences, 193
American Academy of Family Physicians, 295n9
American Bar Foundation, 295n8
American Council on Education, 61, 65
American Institute of Physics, 60, 64
Anderson, Nels, 5
Andrews, Frank, 14, 246
anomie, 221–227, 260, 266–267, 268
Atkinson, Robert, 53

Baldwin, Roger, 220, 221
Ball, Donald, 248
Baltes, Paul, 8
Barley, Steven, 6, 297n14
Barron's Educational Series, 61, 65
Bayer, Alan, 246
Beach, Steven, xi
Becker, Howard, 5, 153, 240
Becker, Tony, 296n11
Ben-David, Joseph, 16
Bentley, Richard, 221

Bergman, Lars, 53
Bertaux, Daniel, 53
Bethe, Hans, 56
Bjorklund, Diane, 56
Blackburn, Robert, 14, 79, 220, 221, 301n6
Blau, Peter, 54
Bledstein, Burton, 261
Braxton, John, 246, 254, 300n2
Brint, Steven, 265
Bruer, John, 128, 247
Burawoy, Michael, 53

Cairns, Robert, 53
Capoulie, Jean-Michel, 297n13
careers, 6, 16, 21–22, 217; advancement of knowledge and, 15; expectations and, 218–227, 253, 260–262, 266–267, 268; how institutions shape, 4–7, 15, 26–27, 34, 267–268; in life course perspective, 9; mobility in, 33–34, 37–38; modes of adaptation in, 225–227, 300n1; objective, 6, 57; orderly vs. "chaotic," 300–301n5; paradigmatic, 17; reversals of, 104–112, 116, 166, 191–195, 208; and satisfactions, 48, 130–131, 134, 145–146, 148, 151, 160–161, 167, 191, 209, 210, 215, 218–221, 224–225, 301n6; stratification of, 11, 13; subjective, 3, 6, 221–227; turning points and, 6, 159, 233. *See also* professions; stratification

Carnegie Foundation for the Advancement of Teaching, 32
Caspi, Avshalom, 7
Chandrasekhar, S., 17
Charmaz, Kathy, 52
Chicago School of Sociology, 4–5, 297n14
Chudacoff, Howard, 298n4
Clair, Jeffrey, 9
Clark, Burton, 14, 296n11, 301n6
Clark, Robert, 299–300n1
Clausen, John, 7, 53
Clemente, Frank, 95
codification in science, 254
Coggeshall, Porter, 22, 61, 297n1
Cohler, Bertram, 56
Cole, Jonathan, 13, 95, 128, 243, 247, 254, 298n5
Cole, Stephen, 13, 95, 128, 234, 243, 246, 247, 254, 268, 298n5
consensus in science, 254–256
"cooling-out," 247–252, 260
Corbin, Juliet, 53
Cowley, Malcolm, 295n4
Crane, Diana, 95, 246, 298n2
Cressey, Donald, 5
cumulative advantage/disadvantage, theory of, 12, 13, 14, 223, 238, 261. *See also* recognition; stratification

Dannefer, Dale, 8, 238
Datan, Nancy, 9
Davis, Fred, 5
Denzin, Norman, 53
DiPrete, Thomas, 12
Donovan, Francis, 5
Dorfman, Lorraine, 299n1
Dubin, Robert, 155
Duncan, Otis Dudley, 54
Dunham, Edgar Alden, 265
Durkheim, Emile, xii, 25, 53, 221, 222, 223, 227, 237, 239
Dutton, Jeffrey, 246
Dyer, Thomas, xi
Dyson, Freeman, 56

Eiduson, Bernice, 228
Eihrich, Gregory, 12
Eisenhart, Margaret, 128
Elder, Glen H., Jr., 7, 8, 9, 53
Erikson, Erik, 8

Faulkner, Robert, 248, 250
Faulkner, William, 1, 295n4
Feynman, Richard, 202
Finkel, Elizabeth, 128
Finkelstein, Martin, 14, 34, 74, 79, 295n10, 301n6
Finlay, William, xi
Finnegan, D. E., 265
Flattau, Pamela Ebert, 22, 60, 64, 297n1, 298n3
Foner, Anne, 10
Foote, Elizabeth, xii
Fox, Mary Frank, 95, 128, 247
Friedson, Eliot, 5
Fry, Christine, 298n4
funding for research: concerns about, 96–97, 109, 135, 256–258; government levels of, 72; variation in departmental, 58–66

Gamson, Z., 265
Gaston, Jerry, 13, 95
Geer, Blanche, 5
Geertz, Clifford, 25
Giele, Janet, 53
Glaser, Barney, 7, 14, 52, 53, 225, 300n4
Glashow, Sheldon, 17
Glueck, Eleanor, 54
Glueck, Sheldon, 54
Goffman, Erving, 5, 247, 249, 251
Goldberger, Marvin, 22, 60, 64, 297n1, 298n3
Goldner, Fred, 248
Goode, William, 14, 228
Gouldner, Alvin, 152, 231, 232, 300n3
Grant, Linda, xii
Gregorio, David, 247
Gustin, Bernard, 14

Habenstein, Robert, 5
Haber, Samuel, 261
Hagstrom, Warren, 95, 222, 223, 224, 225, 300n1
Hammond, P. Brett, 299–300n1
Hanneman, Robert, 265
Hargens, Lowell, 95, 233, 254
Harper, Douglas, 6
Hayner, Norman, 5
Hearn, James, xi
Henderson, B. B., 265
Herbener, Ellen, 7
Hermanowicz, Joseph, 3, 21, 24, 25, 27, 28, 31, 56, 60, 61, 83, 128–129, 143, 259, 262, 301n6
Hernstein, Richard, 261
Hiles, Sara Shipley, 268
Holton, Gerald, 128, 129, 247
Hostetler, Andrew, 56
Hughes, Everett, 4, 5, 6, 7, 13, 15, 57, 233, 234, 297n13

identity: See self-identity
International Labour Office, 295n7

Jones, Lyle, 22, 61, 297n1

Kagan, Jerome, 53
Kalleberg, Arne, 224, 225
Kane, W. D., 265
Karp, David, 9
Kelly-Wilson, Lisa, 254
Kohli, Martin, 53, 298n4
Kohn, Melvin, 224, 297n16
Krauze, Tad, 13, 247

LaFollette, Marcel, 300n2
Lashbrook, Jeff, 298n4
Laub, John, 54
Lawrence, Barbara, 227, 298n4
Lawrence, Janet, xi, 14, 79, 221, 301n6
Lawson, Helene, 6
Lehman, Harvey, 246
Levin, Sharon, 246

Levinson, Daniel, 8, 53, 87
Lewis, Lionel, 247
life course, 2, 4, 7–11; age grades and, 9; "fallacy of the," 10; social control of the, 240–252; stage theories and, 8–9
Light, Donald, 296n11
Linde, Charlotte, 54
Lindzey, Gardner, 22, 61, 297n1
Long, J. Scott, 13, 95, 246, 247, 298n2, 298n5
longitudinal analysis, 3, 7, 8, 10–11, 19, 21–22, 39–40, 45–56, 296–297n12, 297n15
Loscocco, Karyn, 225
Lowe, John, 9

Maher, Brendan, 22, 60, 64, 297n1, 298n3
Martin, Joanne, 56
"Matthew Effect," 12–13
McClelland, David, 251, 261
McCormmach, Russell, 17
McGinnis, Robert, 95, 246, 298n5
McGovern, Timothy, xii
Merton, Robert K., xii, 12, 13, 15, 30, 95, 221, 225, 227, 229, 234, 238, 247, 251, 252, 254, 298n4, 299n2, 300n1, 300n3
Meyer, John, 298n4
"mission creep," 265–266
Mitchell, Douglas, xii
Mitchell, Richard, 52
mobility: in careers, 33–34, 37–38; perceived lack of, 117; reasons for, 35–37, 101–102; among universities, 33–37
Monastersky, Richard, 268
Moore, Joan, 9
Morphew, Christopher, 265
Morris, Libby, xi
Morse, Janice, 30
Mortimer, Jeylan, 8
Morus, Iwan Rhys, 17
Murray, Charles, 261

National Academy of Sciences, 110, 152, 189, 193, 194, 202
National Center for Education Statistics, 301n6
National Opinion Research Center, 301n6
National Research Council, 22, 27, 28, 128, 247, 297n1, 297n1, 298n3
National Science Board, 73
National Science Foundation, 107
Neave, Guy, 265
Nesselroade, John, 8
Neugarten, Bernice, 9
New York Times, 45, 89
Nobel Prize, 87, 88, 160, 165, 193, 202
norms of science, 95, 166, 299n2

"ontogenetic fallacy," 8
Organization for Economic Cooperation and Development, 295n7

Park, Robert, 5
Parkes, Colin, 54
Parsons, Talcott, 16
particularism in science. *See* norms of science
Paul, Charles, 16, 261
Pawluch, Dorothy, 297n12
Pelz, Donald, 14, 246
Phelps, Erin, 297n15, 299n7
physics, 16–19, 253–259
Platt, Gerald, 16
Preston, Anne, 128
professions: higher education treatment of, 296n11; internal differentiation of, 3–4, 21–22, 31–33, 104, 136, 260; science and, 16, 217, 243–245, 251–252, 268; work hours in, 2, 295n8, 295n9, 295n10. *See also* stratification; careers
publication: concern about, 90–91; perceived politics in, 124; quality and quantity of, 78–82, 116, 139–142

Ragin, Charles, 54
recognition, 12, 129, 172, 213, 219; centrality of, to science, 12, 30, 84, 90, 229, 249, 252; equity in, 153, 159–160, 176, 197–198, 202–203, 215–216; perception of, 95, 106, 126, 171, 196, 214, 234, 241–245; rewards and, 12, 223
reference groups, 227–239
Renzulli, Linda, xii
Reskin, Barbara, 95, 246, 247
retirement, 44, 49, 178, 215, 299–300n1; life in, 167, 168, 187–188, 189, 192, 200, 201, 206; thoughts of, 94, 147–148, 162, 187
Reynolds, Jeremy, 295n4
Rhoades, Gary, 217
Riddle, Mark, 265
Riley, Matilda White, 10
Ritti, R. R., 248
Roe, Anne, 297n12
Roman, Paul, xi
Rosenberg, Morris, 297n16
Roth, Julius, 298n4
Rubin, David, 56
Rubin, Herbert, 30
Rubin, Irene, 30
Rubin, Leonard, 95
Ruch, Richard, 265
Ruppersburg, Hugh, 295n4
Ruscio, Kenneth, 296n11

Saldana, Johnny, 51
Sampson, Robert, 54
Savola, Kristen, 297n15, 299n7
Schooler, Carmi, 297n16
Schuster, Jack, xi, 14, 34, 74, 79, 295n10, 301n6
Schwartz, Barry, xii
Science, 45
Science Citation Index, 299n6
Seal, Robert, 14, 74, 79, 295n10
selection effects, 259–260
self-identity, 7

Settersten, Richard, 298n4
Shaffir, William, 297n12
Shauman, Kimberlee, 128, 247
Shaw, Clifford, 5
Sheehy, Gail, 8
Shibutani, Tamotsu, 56, 228, 232
Simmons, Roberta, 8
Simon, Gary, 254
Simon, Rita James, 261
Slaughter, Sheila, xi
Sonnert, Gerhard, 128, 129, 247
Spenner, Kenneth, 297n16
Stars Are Not Enough: Scientists—Their Passions and Professions, 3, 21, 24
Stebbins, Robert, 6
Stephan, Paula, 246
Stewart, John, 13, 246
stratification: and advancement of social institutions, 15; centrality of research in, 15; in publication, 79–82, 139–142, 179–182; reward systems and, 11, 13–14, 95, 134, 163–164; in science, 11–16, 235; of scientists by rank, 40–43; of universities, 25–29
Strauss, Anselm, 5, 7, 52, 53
Stryker, Sheldon, 56
Szilard, Leo, 56

teaching: changed perceptions of, 100, 101, 102, 108, 119, 154, 155, 186, 206; course loads and, 58–65; vs. research, 163, 239, 262
Thomas, W. I., 150
Thorgerson, Eric, xii
TIAA-CREF Institute, xi
Trowler, Paul, 296n11

United States Department of Education, 296n12

United States Department of Labor, 1, 295n1, 295n2, 295n5
universalism in science. *See* norms of science
universities: as academic worlds, 22–27; change in, 57–75; classification of, 31–33, 297–298n1; communitarian, 22, 23–24, 235; elite, 22–23, 235; mobility among, 33–37, 298n5; pluralist, 22–23, 235–237; scientists' views of, 85–86, 121–122, 135–136, 145, 164, 166–167, 175, 185, 199, 205, 211, 219

Vaillant, George, 53
Van Maanen, John, 6

Wanner, Richard, 247
Waring, Joan, 10
Weber, Max, 167
Weiss, Robert, 53, 54, 299n1
Wells, Edward, 56
White, Robert, 54
Wilensky, Harold, 300n5
Wilson, Logan, 296–297n12
women scientists, 38–39, 125–127, 128–129, 162, 188–189, 195–196, 258–259
work hours, 1–2, 295n1, 295n2, 295n3, 295n5, 295n6, 295n7, 295n8

Xie, Yu, 128, 247

Yoels, William, 9
Young, Copeland, 297n15, 299n7

Zuckerman, Harriet, 11, 12, 14, 15, 95, 128, 227, 234, 238, 243, 246, 247, 254, 266